# An Introduction to Analytical Atomic Spectrometry

# ERRATA

## Introduction to Atomic Spectroscopy
### L. C. Ebdon *et al.*

page 134: Equation (5.7) and the paragraph following it should read as follows—

$$R_m = \frac{A_x C_x W_x M_s + A_s C_s W_s M_x}{B_x C_x W_x M_s + B_s C_s W_s M_x} \tag{5.7}$$

where: $R_m$ = observed isotope ratio of A to B, $A_x$ = atom fraction of isotope A in sample, $B_x$ = atom fraction of isotope B in sample, $A_s$ = atom fraction of isotope A in spike, $B_s$ = atom fraction of isotope B in spike, $W_x$ = weight of sample, $W_s$ = weight of spike, $C_x$ = concentration of element in sample and $C_s$ = concentration of element in spike, and $M_x$ and $M_s$ = molar mass of element in sample and spike respectively.

Equation (5.8) and the paragraph following it should read as follows—

$$C_x = \frac{C_s W_s M_x}{W_x M_s} \times \frac{A_s - R_m B_s}{R_m B_x - A_x} \tag{5.8}$$

The concentration of any isotope, for example isotope A, can be calculated from the relationship $C_x A_x M_A / M_x$, where $M_A$ is the molar mass of isotope A.

page 178: Equation (B.2) and the paragraph following it should read as follows—

$$R = \frac{A_x C_x W_x M_s + A_s C_s W_s M_x}{B_x C_x W_x M_s + B_s C_s W_s M_x} \tag{B.2}$$

where $R$ = isotope ratio of A to B, $A_x$ = atom fraction of isotope A ($^{106}Cd$) in sample, $B_x$ = atom fraction of isotope B ($^{111}Cd$) in sample, $A_s$ = atom fraction of isotope A ($^{106}Cd$) in spike, $B_s$ = atom fraction of isotope B ($^{111}Cd$) in spike, $W_x$ = weight of sample, $W_s$ = weight of spike, $C_x$ = concentration of element in sample and $C_s$ = concentration of element in spike, and $M_x$ and $M_s$ = molar mass of element in sample and spike respectively.

Equations (B.3) and (B.4) should read as follows—

$$W_s = \frac{W_x C_x M_s (A_x - R B_x)}{C_s M_x (R B_s - A_s)} \tag{B.3}$$

$$C_x = \frac{C_s W_s M_s}{W_x M_x} \times \frac{A_s - R B_s}{R B_x - A_x} \tag{B.4}$$

# An Introduction to Analytical Atomic Spectrometry

**Contributing Authors:**

# L. Ebdon
# E.H. Evans
# A.S. Fisher
# S.J. Hill

*University of Plymouth, UK*

**Edited by:**

# E.H. Evans

JOHN WILEY & SONS

Chichester • New York • Weinheim • Brisbane • Singapore • Toronto

*Other Wiley Editorial Offices*

John Wiley & Sons, Inc., 605 Third Avenue,
New York, NY 10158-0012, USA

WILEY-VCH Verlag GmbH, Pappelallee 3,
D-69469 Weinheim, Germany

Jacaranda Wiley Ltd, 33 Park Road, Milton,
Queensland 4064, Australia

John Wiley & Sons (Asia) Pte Ltd, 2 Clementi Loop #02-01,
Jin Xing Distripark, Singapore 129809

John Wiley & Sons (Canada) Ltd, 22 Worcester Road,
Rexdale, Ontario M9W 1L1, Canada

*Library of Congress Cataloging-in-Publication Data*

An introduction to analytical atomic spectroscopy / contributing
authors, L. Ebdon ... [et al.] ; edited by E.H. Evans.
    p.  cm.
    Includes bibliographical references and index.
    ISBN 0-471-97417-X (alk. paper). — ISBN 0-471-97418-8 (pbk.:
alk. paper)
    1. Atomic spectroscopy.   I. Ebdon, L.   II. Evans, E. Hywel.
QD96.A8158   1998
543'.0873 — dc21                                              97-31697
                                                                CIP

*British Library Cataloguing in Publication Data*

A catalogue record for this book is available from the British Library

ISBN 0 471 97417 X (cloth)
ISBN 0 471 97418 8 (paper)

Typeset in 10/12pt Palatino by Laser Words, Madras, India

# CONTENTS

# PREFACE

This book is based on *An Introduction to Atomic Absorption Spectroscopy* by L. Ebdon, which was published in 1982. Since then there have been a number of significant developments in the field of Atomic Spectrometry: inductively coupled plasma atomic emission spectrometry (ICP-AES) has become an established technique, and is used in most analytical laboratories; the spectacular rise to prominence of inductively coupled plasma mass spectrometry has occurred, with a concomitant increase in the speed and quantity of data production, and the sensitivity of analyses. To reflect these changes we have chosen the more generally applicable title *An Introduction to Analytical Atomic Spectrometry* for this book. While much of the original text from *An Introduction to Atomic Absorption Spectroscopy* has been retained, the chapter on Plasma Atomic Emission Spectrometry has been expanded to reflect the importance of ICP-AES, and a chapter on Inductively Coupled Plasma Mass Spectrometry has been included. A thorough treatment of Flame Atomic Absorption Spectrometry (FAAS) has been retained because a thorough understanding of this technique will form the basis of understanding in the whole field of analytical atomic spectrometry. Just as importantly, FAAS is available in most teaching laboratories, whereas ICP-AES and ICP-MS are not.

The rationale of this book remains the same as that of its forerunner. The book is intended to complement undergraduate and postgraduate courses in analytical chemistry, and to aid in the continuing professional development of analytical chemists in the workplace. The problems of release from work to engage in training are even more acute now than they were in 1982, despite the even greater necessity for lifelong learning and continuous upgrading of skills. Even in full-time education the situation has changed. The number of students studying for first and second degrees has increased, and mature students are returning to education in greater numbers than ever before, hence distance and self-learning have become an even more vital component in any course of study.

Keywords are highlighted throughout the text, and there are self-assessment questions at intervals throughout. Chapter 1 is a brief overview of theory and instrumentation. A short treatment of good laboratory practice and sample preparation is included. No amount of words can do justice to these issues, so the discussion is limited to the main points, with the onus being on the tutor to impress upon the student the importance of quality assurance in the practical environment of the laboratory itself! Flame and electrothermal atomic absorption spectrometry are dealt with in Chapters 2 and 3, respectively, revised to take account of recent developments. Plasma emission spectrometry is dealt with in Chapter 4, with centre stage going to the inductively coupled plasma. Inductively coupled plasma mass spectrometry is the subject of Chapter 5. Two short Chapters, 6 and 7, then deal with atomic fluorescence spectrometry and special sample introduction methods. In each of the chapters there are sections on theory, instrumentation, interferences and applications. Several appendices contain revision questions, practical and laboratory exercises and a bibliography.

The book can be used as a self-learning text but it is primarily meant to complement a lecture or distance learning course, and is indeed used in this capacity at Plymouth for undergraduate and postgraduate lectures, and for short courses. Basic theory is included because this is vital to the understanding of the subject; however, excessive theoretical discourse has been avoided, and the emphasis is firmly on the practical aspects of analytical atomic spectrometry.

E. Hywel Evans

Plymouth
July 1997

# ACKNOWLEDGEMENTS

Thanks to all those, colleagues and students, who involved themselves in discussion over many years — their essence is distilled in this book. Also, many thanks to Siân for proof-reading and typing, and to Gavin for practical B.6.

# 1 OVERVIEW OF ANALYTICAL ATOMIC SPECTROMETRY

## 1.1 HISTORICAL

### 1.1.1 Optical spectroscopy

Spectroscopy is generally considered to have started in 1666, with **Newton**'s discovery of the solar spectrum. **Wollaston** repeated Newton's experiment and in 1802 reported that the sun's spectrum was intersected by a number of dark lines. **Fraunhofer** investigated these lines — Fraunhofer lines — further and, in 1823, was able to determine their wavelengths.

Early workers had noted the colours imparted to diffusion flames of alcohol by metallic salts, but detailed study of these colours awaited the development of the premixed air–coal gas flame by **Bunsen**. In 1859, **Kirchhoff** showed that these colours arose from line spectra due to elements and not compounds. He also showed that their wavelengths corresponded to those of the Fraunhofer lines. Kirchhoff and Fraunhofer had been observing atomic emission and atomic absorption, respectively.

Atomic absorption spectroscopy (AAS), atomic emission spectroscopy (AES) and later atomic fluorescence spectroscopy (AFS) then became more associated with an exciting period in astronomy and fundamental atomic physics. **Atomic emission spectroscopy** was the first to re-enter the field of analytical chemistry, initially in arc and spark spectrography and then through the work of **Lunegardh**, who in 1928 demonstrated AES in an air–acetylene flame using a pneumatic nebulizer. He applied this system to agricultural analysis. However, the technique was relatively neglected until the development of the inductively coupled plasma as an atom cell, by **Greenfield** in the UK and **Fassel** in the USA, which overcame many of the problems associated with flames, arcs and sparks. The term emission spectroscopy is applied to the measurement of light emitted from flames or plasmas by chemical species after the absorption of energy as heat or as chemical energy (i.e. chemiluminescence). If only the emission from atoms is observed, the term **atomic emission spectroscopy** is preferred.

**Atomic absorption spectroscopy** is the term used when the radiation absorbed by atoms is measured. The application of AAS to analytical problems was considerably delayed because of the apparent need for very high resolution to make quantitative measurements. In 1953, **Walsh** brilliantly overcame this obstacle by the use of a line source, an idea pursued independently by Alkemade, his work being published in 1955.

The re-emission of radiation from atoms which have absorbed light is termed **atomic fluorescence**. In 1962, **Alkemade** was the first to suggest that AFS had analytical potential, which was demonstrated in 1964 by **Winefordner**.

These three types of spectroscopy are summarized in Fig. 1.1. The horizontal lines represent different energy levels in an atom. $E_0$ is the term

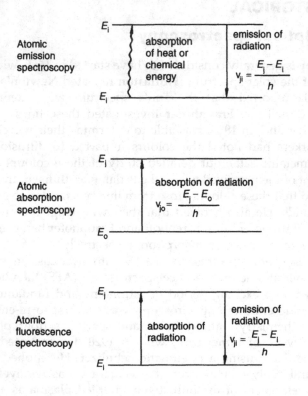

**Figure 1.1** Summary of AES, AAS and AFS.

used for the lowest energy level, which is referred to as the ground state and therefore all practical absorption measurements originate from atoms in the ground state, as do virtually all practical fluorescence measurements. $E_i$ and $E_j$ refer to other energy levels, $E_j$ being higher (greater energy) than $E_i$. A solid vertical line refers to a transition involving the absorption or emission of radiation as energy. The wavy line refers to a non-radiative transition.

The energy of the radiation absorbed or emitted is quantized according to **Planck's equation** (Eqn. 1.1). These quanta are known as photons, the energy of which is proportional to the frequency of the radiation.

$$E = h\nu = hc/\lambda \tag{1.1}$$

where $h$ = Planck's constant $(6.62 \times 10^{-34}$ J s$)$, $\nu$ = frequency, $c$ = velocity of light $(3 \times 10^8$ m s$^{-1})$ and $\lambda$ = wavelength (m).

## 1.1.2 Mass spectrometry

After the existence of different isotopic forms of the elements was first demonstrated by **Thompson**, the technique of mass spectrometry is considered to have begun with **Aston** and **Dempster**, who reported the accurate measurements of ionic masses and abundances in 1918–19. From these beginnings, using magnetic and electric fields to separate ions of different mass, mass spectrometry has grown into a major technique for the analysis of organic and inorganic compounds and elements. The development of instruments based on quadrupole, time-of-flight, and ion trap mass analysers has taken the technique from the research laboratory into everyday use as an analytical instrument.

In **magnetic/electric sector** mass analysers, ions are **deflected** in a magnetic field, the extent of deflection depending on their **mass-to-charge ratio** $(m/z)$. If all the ions have the same charge then the deflection is dependent on their relative masses, hence a separation can be effected. In **quadrupole** mass analysers the ions are subjected to a radiofrequency (RF) field, which is controlled so that only one particular $m/z$ can pass through it. Hence the quadrupole acts like a **mass filter**, and the field can be varied so that ions of consecutively higher $m/z$ pass through sequentially. An **ion trap** acts in a similar way except that the ions are first trapped inside it, then let out sequentially. **Time-of-flight mass** analysers are essentially long tubes along which the ions pass. Ions of low mass have a higher velocity than ions of higher mass so, if a pulse of ions is introduced into one end of the tube, the light ions will arrive at the other end before the heavy ions, thereby effecting a separation.

Mass analysis is a relatively simple technique, with the number of ions detected being directly proportional to the number of ions introduced into the mass spectrometer from the **ion source**. In atomic mass spectrometry the ion source produces **atomic ions** (rather than the molecular ions formed for qualitative organic analysis) which are proportional to the concentration of the element in the original sample. It was **Gray** who first recognized that the inductively coupled plasma would make an ideal ion source for atomic mass spectrometry and, in parallel with **Fassel** and **Houk**, and **Douglas** and **French** developed the ion sampling interface necessary to couple an atmospheric pressure plasma with a mass spectrometer under vacuum.

## 1.2  BASIC INSTRUMENTATION

### 1.2.1  Optical spectroscopy

Figure 1.2 shows the basic instrumentation necessary for each technique. At this stage, we shall define the component where the atoms are produced and viewed as the 'atom cell'. Much of what follows will explain what we mean by this term. In atomic emission spectroscopy, the atoms are excited in the atom cell also, but for atomic absorption and atomic fluorescence spectroscopy, an external light source is used to excite the ground-state atoms. In atomic absorption spectroscopy, the source is viewed directly and the attenuation of radiation measured. In atomic fluorescence spectroscopy, the source is not viewed directly, but the re-emittance of radiation is measured.

Current instrumentation usually uses a diffraction grating as the dispersive element and a photomultiplier as the detector, although solid-state detectors are becoming more widespread.

Historically, data collection and manipulation were effected by means of analogue meters, chart recorders, digital displays and paper print-outs. However, the advent of the microcomputer now allows data to be stored electronically, calibrations performed and concentrations calculated and reported on a user-defined form.

### 1.2.2  Mass spectrometry

Figure 1.2 shows the basic instrumentation for atomic mass spectrometry. The component where the ions are produced and sampled from is the 'ion source'. Unlike optical spectroscopy, the ion sampling interface is in intimate contact with the ion source because the ions must be extracted into the vacuum conditions of the mass spectrometer. The ions are separated with respect to mass by the mass analyser, usually a quadrupole, and literally counted by means of an electron multiplier detector. The ion signal for each

### Atomic Emission

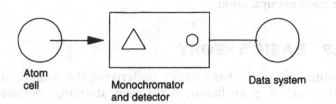

### Atomic Absorption

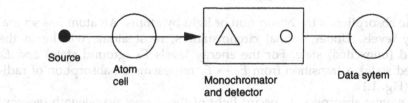

### Atomic Fluorescence

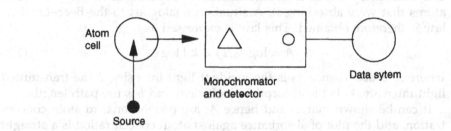

### Atomic Mass Spectrometry

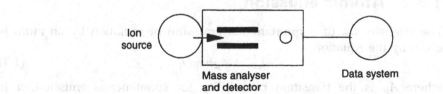

**Figure 1.2**   Basic instrumentation systems used in analytical atomic spectrometry.

mass is stored in a multi-channel analyser and output to a microcomputer for data manipulation.

## 1.3  BASIC THEORY

A summary of the basic theory underlying the main atomic spectrometric techniques is given below. For a more thorough discussion, refer to the individual chapters.

### 1.3.1  Atomic absorption

Atomic absorption is the absorption of light by atoms. An atom has several energy levels. Under normal circumstances, most atoms will be in the ground (unexcited) state. For the energy levels $E_0$ (ground state) and $E_j$ (excited state), a transition from $E_0 \rightarrow E_j$ represents an absorption of radiation (Fig. 1.1).

For atomic absorption to occur, light of the correct wavelength (energy) is absorbed by ground-state electrons, promoting them to a higher, excited state. The intensity of the light leaving the analytes is therefore diminished. The amount by which it is diminished is proportional to the number of atoms that were absorbing it. A situation analogous to the Beer–Lambert law is therefore obtained. This law is expressed as

$$A = \log(I_0/I) = k_v l \log e \qquad (1.2)$$

where $A$ is absorbance, $I_0$ is the incident light intensity, $I$ the transmitted light intensity, $k_v$ is the absorption coefficient and $l$ is the path length.

It can be shown that $k_v$ and hence $A$, are proportional to atom concentration, and the plot of absorbance against atom concentration is a **straight line**.

### 1.3.2  Atomic emission

The intensity $I_{em}$ of a **spontaneous emission** of radiation by an atom is given by the equation

$$I_{em} = A_{ji} h v_{ji} N_j \qquad (1.3)$$

where $A_{ji}$ is the transition probability for spontaneous emission, $h$ is Planck's constant, $v_{ji}$ is the frequency of radiation and $N_j$ the number of atoms in the excited state.

It can be shown (see Chapter 4) that $N_j$, and hence $I_{em}$, are proportional to the atom concentration, and for low concentrations the plot of emission intensity against atom concentration is a **straight line**.

## 1.3.3  Atomic fluorescence

In atomic fluorescence spectroscopy an intense **excitation source** is focused on to the atom cell. The atoms are excited then re-emit radiation, in all directions, when they return to the ground state. The radiation passes to a detector usually positioned at right-angles to the incident light. At low concentrations, the intensity of fluorescence is governed by the following relationship:

$$I_f = k \Phi I_0 C \tag{1.4}$$

where $I_f$ is the intensity of fluorescent radiation, $C$ is the concentration of atoms, $k$ is a constant, $I_0$ is the intensity of the source at the absorption line wavelength and $\Phi$ is the **quantum efficiency** for the fluorescent process (defined as the ratio of the number of atoms which fluoresce from the excited state to the number of atoms which undergo excitation to the same excited state from the ground state in unit time).

The intensity of fluorescence is proportional to the concentration of atoms, and hence the concentration of the element in the sample, so a plot of concentration against fluorescence will yield a **straight line**.

There are several different types of atomic fluorescence as follows:

(i)   Resonance fluorescence.
(ii)  Direct line fluorescence.
(iii) Stepwise line fluorescence.
(iv)  Thermally assisted fluorescence.

These are described in more detail in Chapter 6. Resonance fluorescence, i.e. the excitation and emission are at the same wavelength, is most widely used. The others have very limited use analytically.

## 1.3.4  Atomic mass spectrometry

The **degree of ionization** of an atom is given by the **Saha equation**:

$$\frac{n_i n_e}{n_a} = 2\frac{Z_i}{Z_a}\left(2\pi mk \frac{T}{h^2}\right)^{3/2} \exp(-E_i/kT) \tag{1.5}$$

where $n_i$, $n_e$ and $n_a$ are the number densities of the ions, free electrons and atoms, respectively, $Z_i$ and $Z_a$ are the ionic and atomic partition functions, respectively, $m$ is the electron mass, $k$ is the Boltzmann constant, $T$ is the temperature, $h$ is Planck's constant and $E_i$ is the first ionization energy.

In atomic mass spectrometry, the rate of production of ions is measured directly. This is proportional to the concentration of ions, and hence atoms. A plot of ion count rate against atom concentration will therefore yield a **straight line**.

## 1.4  PRACTICE

### 1.4.1  Calibration and analysis

Spectroscopic techniques require **calibration** with standards of known analyte concentration. Atomic spectrometry is sufficiently specific for a simple solution of a salt of the analyte in dilute acid to be used, although it is a wise precaution to **buffer** the standards with any salt which occurs in large concentration in the sample solution, e.g. 500 μg ml⁻¹ or above. **Calibration curves** can be obtained by plotting **absorbance** (for AAS), **emission signal** (for AES), **fluorescence signal** (for AFS) or **ion count rate** (for MS) as the dependent variable against concentration as the independent variable. Often the calibration curve will **bend** towards the concentration axis at higher concentrations, as shown in Fig. 1.3. In AAS this is caused by **stray** or **unabsorbable** light, in AES and AFS by **self-absorption** and in MS by **detector overload**. As the slope decreases, so will **precision** and it is preferable to work on the **linear portion** of the calibration known as the **working curve**. The best results are obtained when the standards are introduced first in ascending order of concentration, and then in descending order, each time 'bracketing' the samples with standards of immediately lower then higher concentration when ascending, and the reverse when descending. Modern instruments will normally have computer software for effectively performing the calibration. Samples should **not lie off** the

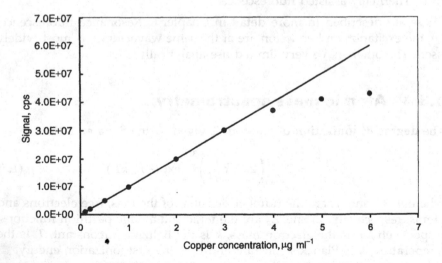

**Figure 1.3** Typical calibration curve obtained in atomic spectrometry. At high concentrations the curve will bend towards the concentration axis; for explanation, see text.

calibration curve, i.e. it must never be more concentrated than the strongest standard and preferably not more dilute than the weakest.

The method of **standard additions** is a useful procedure for checking the accuracy of a determination and overcoming interferences when the composition of the sample is unknown. It should be noted that the method cannot be used to correct for **spectral interferences** and **background changes**. At least three aliquots of the sample are taken. One is left untreated; to the others **known additions** of the analyte are made. The additions should preferably be about $0.5x$, $x$ and $2x$, where $x$ is the concentration of the unknown. It should also be noted that the volume of the addition should be negligible in comparison with the sample solution. This is to prevent dilution effects

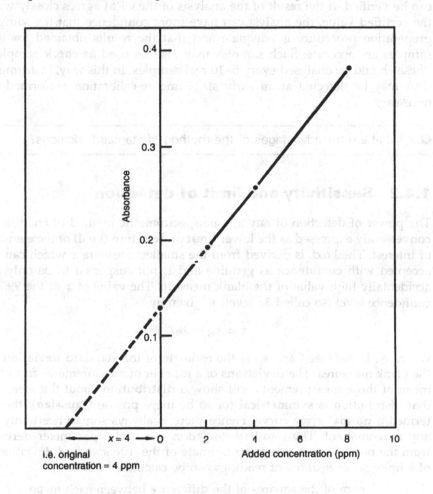

**Figure 1.4** Method of standard additions.

which would render the standard addition process invalid. The solutions are then aspirated and the curve shown in Fig. 1.4 is plotted. The curve is **extrapolated** back until it crosses the $x$-axis, giving the concentration in the unknown. A standard addition curve **parallel** to the calibration curve is indicative (but not conclusive) of the absence of interference.

Accuracy of analyses should be checked using **certified reference materials** (CRMs). These are materials ranging from botanical and biological to environmental and metallic samples that have been analysed by numerous laboratories using several independent techniques. As a result, 'agreed' values of the sample's elemental composition are produced. Therefore, by matching a CRM with the sample to be analysed, the validity of the analysis can be verified. If the result of the analysis of the CRM agrees closely with the **certified** value, the analyst can have more confidence that the sample preparation procedure is adequate and that the results obtained for the samples are accurate. Such samples may also be used as **check** samples. These should be analysed every 5–10 real samples. In this way, instrument drift may be detected at an early stage and re-calibration performed as necessary.

---

**Q.**  What are the advantages of the method of standard additions?

---

## 1.4.2  Sensitivity and limit of detection

The power of detection of any atomic spectrometric method of analysis is conveniently expressed as the lower **limit of detection (l.o.d)** of the element of interest. The l.o.d. is derived from the smallest measure $x$ which can be accepted with **confidence** as genuine and is not suspected to be only an accidentally high value of the blank measure. The value of $x$ at the 99.7% confidence level (so called **3s level**) is given by

$$x = x_{bl} + 3s_{bl}$$

where $x_{bl}$ is the mean and $s_{bl}$ is the estimate of the **standard deviation** of the blank measures. The **deviations** of a number of measurements from the mean of those measurements will show a **distribution** about the mean. If that distribution is **symmetrical** (or to be more precise **Gaussian**), this is termed a **normal error curve**. Hence there is always some uncertainty in any measurement. The standard deviation is a useful parameter derived from the normal error curve. An estimate of the true standard deviation $s$ of a finite set of $n$ different readings can be calculated from

$$s = \frac{\text{sum of the squares of the difference between each mean}}{n-1}$$

Statistical theory tells us that, provided sufficient readings are taken, 68.3% of the actual readings lie within the standard deviation of the mean, and that the mean ±2s and the mean ±3s will contain 95.5% and 99.7% of the readings, respectively. Hence there is only a 0.3% chance that, if the readings are larger than the mean of the blank readings by three times the standard deviation, this is merely due to an unusually high blank reading. Thus, the limit of detection may be defined 'as that quantity of element which gives rise to a reading equal to three times the standard deviation of a series of at least 10 determinations at or near the blank level'. This assumes a 'normal' distribution of errors, and may consequently result in more or less optimistic values.

The limit of detection is a useful figure which takes into account the stability of the total instrumental system. It may vary from instrument to instrument and even from day to day as, for example, mains-borne noise varies. Thus, for atomic absorption techniques, spectroscopists often also talk about the **characteristic concentration** (often erroneously referred to as the **sensitivity** — erroneously as it is the reciprocal of the sensitivity) for **1% absorption**, i.e. that concentration of the element which gives rise to **0.0044 absorbance units**. This can easily be read off the calibration curve. The characteristic concentration is dependent on such factors as the atomization efficiency and flame system, and is independent of noise. Both this figure and the limit of detection give different, but useful, information about **instrumental performance**.

---

**Q.** Define the limit of detection and characteristic concentration?

---

**Q.** What information is given by the limit of detection, and how does this differ from that given by the characteristic concentration?

---

# 1.5  INTERFERENCES AND ERRORS

## 1.5.1  Interferences

Interference is defined as an effect causing a **systematic deviation** in the measurement of the signal when a sample is nebulized, as compared with the measure that would be obtained for a solution of equal analyte concentration in the same solvent, but in the absence of **concomitants**. The interference may be due to a particular concomitant or to the combined effect of several concomitants. A concomitant causing an interference is called an **interferent**. Interference only causes an **error** if not adequately corrected for during an analysis. Uncorrected interferences may lead to either **enhancements** or **depressions**. Additionally, errors may arise in analytical methods in other ways, e.g. in **sample pretreatment** via the

**operators** and through the **instrumentation**. We shall deal first with errors and then look at interferences in some detail. Interferences specific to individual techniques are discussed in more detail in the relevant chapters.

### 1.5.1.1 Sample pre-treatment errors

Obviously, the **accuracy** of an analysis depends critically on how **representative** the sample is of the material from which it is taken. The more **heterogeneous** the material, the greater the care must be taken with sampling. The analytical methods described in this book can typically be used on small samples (100 mg of solid or 10 cm$^3$ of liquid), and this again heightens the problem. Readers are referred to a general analytical text for details on sampling, but it should be stressed that if either the concentration of the analyte in the sample does not represent that in the **bulk material**, or the concentration of the analyte in the solution at the time it is presented to the instrument has changed, the resultant error is likely to be greater than any other error discussed here. Regrettably, the supreme importance of these points is not always recognized.

Usually, samples are presented for analysis as liquids. Thus, solid samples must be **dissolved**. **Analytical** or **ultra-high-purity grade reagents** must be used for dissolution to prevent **contamination** at trace levels. Certain volatile metals (e.g. cadmium, lead and zinc) may be lost when **dry ashing**, and volatile chlorides (e.g. arsenic and chromium) lost upon **wet digestion**. It is particularly easy to lose mercury during sample preparation. Appropriate steps must be taken in the choice of method of dissolution, acids and conditions (e.g. whether to use reflux conditions) to prevent such losses.

Another method that has become increasingly popular is **microwave digestion**. Sample may be placed in a PTFE bomb and a suitable digestion mixture of acids added. The bomb is then placed in a microwave oven and exposed to microwave radiation until dissolution is complete. This technique has the advantage that digestion may be accomplished within a few minutes. Some bombs have a pressure release valve. These valves become necessary when oxidizing acids, e.g nitric acid, that produce large quantities of fumes are used. Other bombs do not have these valves, and care must be taken that dangerous or damaging explosions do not occur.

Another method of bomb dissolution involves placing the sample and digestion mixture in a sealed PTFE bomb and then encasing this in a stainless-steel jacket. This may then be placed in a conventional oven for a period of several hours. This technique, although cheaper, takes substantially longer.

Trace metals may be lost by **adsorption** on precipitates, such as the silica formed on digestion using oxidizing acids. This possibility should be investigated (e.g. by recovery tests).

**Glassware** may give rise to further errors. For trace level determinations contamination and losses can occur through (a) surface desorption or leaching and (b) adsorption on surfaces. To avoid contamination, all laboratory glassware should be **washed** with a detergent and thoroughly rinsed, **acid washed** by soaking overnight in 10% v/v nitric acid (sp. gr. 1.41), then rinsed with deionized, distilled water (DDW) and allowed to **equilibrate** in DDW overnight. Disposable plastic-ware such as centrifuge tubes and pipette tips should be similarly treated unless they can be shown to be metal-free. Contamination is a particular problem for analysis by ICP-MS owing to its high **sensitivity**. Some elements (e.g. Al, Pb, Na, Mg, Ca) are ubiquitous in the environment so stringent precautions may be necessary to avoid contamination, such as the use of **clean rooms** and **laminar flow hoods** and the donning of special clothing before entering the laboratory. If ultra-low determinations are to be made then these precautions may be necessary at all times.

**Blanks** should be run for all analyses as a matter of course. Even if high-purity reagents are used, the level of the analyte in the blank may constitute the limiting factor in the analysis, and it may be necessary to **purify** reagents used for dissolution.

Adsorption is also a problem at trace levels. Few solutions below a concentration of 10 μg cm$^{-3}$ can be considered to be **stable** for any length of time. Various **preservatives** to guard against adsorption of metals on to glassware have been reported in the literature. Common precautionary steps are to keep the **acid concentration** high and to use **plastic laboratory ware**.

---

**Q.** Under what conditions is sampling most problematic?

---

**Q.** How can we minimize the possibilities of (i) contamination and (ii) losses by adsorption?

---

## 1.5.2 Operator errors

Experience tells us that no account of possible errors can ever be exhaustive, and some techniques will be more prone to certain types of errors than others.

A frequent source of error is **poor standards**. Besides the obvious error of standards being wrongly made, it should not be forgotten that trace metal standards are **unstable**. Concentrations of 10 μg cm$^{-3}$ and less usually need to be prepared daily. Even standards purchased from commercial suppliers will **age** and this is especially true when chemical changes can be expected in the analyte (e.g. silicon).

Readings should only be taken when the signal has reached **equilibrium**. When a new sample is presented, several seconds will elapse before the system reaches equilibrium. Particular care must be taken when the concentration of the samples or standards changes markedly, especially if the new solution is more **dilute**.

---

**Q.** How can poor standards introduce analytical errors?

---

**Q.** Why should the integration button not be pressed until several seconds after a sample change?

---

## 1.6 APPLICATIONS

Several texts have been published that contain information on applications. In addition, the Atomic Spectrometry Updates published in the *Journal of Analytical Atomic Spectrometry* offer invaluable help in the development of new applications in the laboratory. Nearly all the applications of analytical atomic spectrometry require the sample to be in **solution**. Where possible, samples should be brought into solution to give analyte levels of at least 10 times the limit of detection, known as the **limit of determination**.

### 1.6.1 Clinical, food and organic samples

It is usually necessary to **destroy the organic material** before introducing the sample. Care must be taken to avoid losses of volatile elements (an oxidizing **wet ashing** procedure is preferred for elements such as lead, cadmium and zinc) and **contamination** from reagents. Various mixtures have been used for wet ashing, including hydrogen peroxide–sulphuric acid (1:1) and hydrogen peroxide–nitric acid (1:1) followed by the addition of perchloric acid. Such mixtures should be treated with care owing to the possibility of explosions occurring on the addition of perchloric acid, which can only be used in a stainless-steel fume hood. Beverages must be degassed before spraying. In serum analyses, the protein is often precipitated with trichloroacetic acid before analysis, but only if the analyte is not likely to be coprecipitated. Direct aspiration of diluted samples is to be preferred.

### 1.6.2 Petrochemicals

As organic solvents have different physical characteristics, aqueous standards cannot be used for calibration when determining trace metals in oils or petroleum fractions. The sample can either be ashed or diluted in a common organic solvent (e.g. with 4-methylpentan-2-one), and calibration performed using special **organic standards**. It is not always possible to introduce organic solvents directly, as will be seen.

## 1.6.3 Agricultural samples

In soil analysis, the sample pretreatment varies depending on whether a **total elemental** analysis or an **exchangeable cation** analysis is required. In the former, a silicate analysis method (see below) is appropriate. In the latter, the soil is shaken with an **extractant solution**, e.g. 1 M ammonium acetate, ammonium chloride or disodium EDTA. After filtration, the extractant solution is analysed. Fertilizers and crops can be treated as chemical and food samples, respectively.

## 1.6.4 Waters and effluents

Where the analyte is present in sufficient concentration, it may be determined **directly**. Otherwise it may need to be **concentrated** by evaporation before determination. Methods of concentration include **evaporation, solvent extraction** and preconcentration using **ion exchange** or **chelating resins**.

If information on **total metals** is required, the sample must be acidified before analysis. If information on **dissolved metals** only is required, the sample may be filtered (using a specified pore size) before analysis. Losses may occur however, by adsorption during filtration.

## 1.6.5 Geochemical and mineralogical samples

**Silicate analysis** is not without problems. If measurement of silicon is not required, it may be volatilized off as silicon tetrafluoride, using hydrofluoric acid, although some calcium may be lost as calcium fluoride. Alternatively, sodium carbonate–boric acid **fusions** may be employed. Where possible, final solutions are made up in hydrochloric acid.

## 1.6.6 Metals

Where possible, hydrochloric acid–nitric acid is used to dissolve the sample. The standards may be prepared by dissolving the trace metal in an appropriate solution of the matrix metal [e.g. iron(III) chloride solution for steel].

If possible 1-2% w/v solutions are used. Precautions against interferences may be necessary.

**Q.** Outline methods for the determination of (i) Ca in serum, (ii) Ag in silicate rock and (iii) Mn in steel.

**Q.** What are the advantages of solvent extraction?

# 2 FLAME ATOMIC ABSORPTION SPECTROMETRY

Flame AAS (often abbreviated FAAS) was until recently the most widely used method for trace metal analysis. However, it has now largely been superseded by inductively coupled plasma atomic emission spectrometry (see Chapter 4). It is particularly applicable where the sample is in solution or readily solubilized. It is very simple to use and, as we shall see, remarkably free from interferences. Its growth in popularity has been so rapid that on two occasions, the mid-1960s and the early 1970s, the growth in sales of atomic absorption instruments has exceeded that necessary to ensure that the whole face of the globe would be covered by atomic absorption instruments before the end of the century.

## 2.1 THEORY

Atomic absorption follows an exponential relationship between the intensity $I$ of transmitted light and the absorption path length $l$, which is similar to Lambert's law in molecular spectroscopy:

$$I = I_0 \exp(-k_v l) \tag{2.1}$$

where $I_0$ is the intensity of the incident light beam and $k_v$ is the absorption coefficient at the frequency $v$. In quantitative spectroscopy, absorbance $A$ is defined by

$$A = \log(I_0/I) \tag{2.2}$$

Thus, from Eqn. 2.1, we obtain the linear relationship

$$A = k_v l \log e$$

$$= 0.4343 k_v l \tag{2.3}$$

From classical dispersion theory we can show that $k_v$ is in practical terms proportional to the number of atoms per cubic centimetre in the flame, i.e. $A$ is proportional to analyte concentration.

Atomic absorption corresponds to transitions from low to higher energy states. Therefore, the degree of absorption depends on the population of the lower level. When thermodynamic equilibrium prevails, the population of a given level is determined by Boltzmann's law. As the population of the excited levels is generally very small compared with that of the ground state (that is, the lowest energy state peculiar to the atom), absorption is greatest in lines resulting from transitions from the ground state; these lines are called resonance lines.

Although the phenomenon of atomic absorption has been known since early last century, its **analytical potential** was not exploited until the mid-1950s. The reason for this is simple. Monochromators capable of isolating spectral regions narrower than 0.1 nm are excessively expensive, yet typical atomic absorption lines may often be narrower than 0.002 nm. Figure 2.1 illustrates this, but not to scale! The amount of radiation isolated by the conventional monochromator, and thus viewed by the detector, is not significantly reduced by the **very narrow atomic absorption signal**, even with high concentrations of analyte. Thus, the amount of atomic absorption seen using a **continuum source**, such as is used in molecular absorption spectroscopy, is negligible.

The contribution of Walsh was to replace the continuum source with an **atomic spectral source** (Fig. 2.1). In this case, the monochromator only has to isolate the line of interest from other lines in the lamp (mainly lamp filler gas lines). In Fig. 2.1, we see that the atomic absorption signal exactly overlaps the atomic emission signal from the source and very large reductions in radiation are observed.

Of course, this exact **overlap** is no accident, as atomic absorption and atomic emission lines have the same wavelength. The very narrowness of atomic lines now becomes a positive advantage. The lines being so narrow, the chance of an accidental overlap of an atomic absorption line of one element with an atomic emission line of another is almost negligible. The uniqueness of overlaps in the Walsh method is often known as the 'lock and key effect' and is responsible for the very high **selectivity** enjoyed by atomic absorption spectroscopy.

The best sensitivity is obtained in this method when the **source line is narrower** than the absorption profile of the atoms in the flame. Obviously, the other situation tends towards Fig. 2.1.

In recent years, it has been shown that the construction of atomic absorption spectrometers using **continuum sources** is possible, if somewhat expensive and complicated. So far, commercial manufacturers have not yet produced instruments of this type, which have remained the creations of a

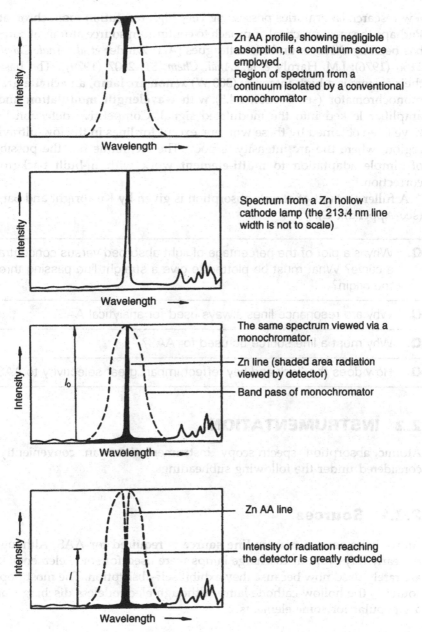

**Figure 2.1** Relative atomic absorption of light from continuum and line sources.

few research laboratories possessing very high resolution monochromators. Perhaps the most practical approach to continuum source atomic absorption has been by O'Haver and his colleagues [A.T. Zander *et al.*, *Anal. Chem.* **48**, 1166 (1976); J.M. Harnly *et al.*, *Anal. Chem.* **51**, 2007 (1979)]. The basis of their system is a high-intensity (300 W) **xenon arc lamp**, an **echelle grating** monochromator (see Section 4.4.5) with **wavelength modulation** and an amplifier locked into the modulated signal. Competitive detection limits have been obtained by these workers, except for lines in the low-ultraviolet region, where the arc intensity is poor. The technique has the possibility of simple adaptation to **multi-element** work with in-built background correction.

A fuller account of atomic absorption is given by Kirkbright and Sargent (see Appendix C).

---

**Q.** Why is a plot of the percentage of light absorbed versus concentration a curve? What must be plotted to give a straight line passing through the origin?

---

**Q.** Why are resonance lines always used for analytical AAS?

---

**Q.** Why must a line source be used for AAS?

---

**Q.** How does the 'lock and key' effect impart great selectivity to AAS?

---

## 2.2  INSTRUMENTATION

Atomic absorption spectroscopy instrumentation can conveniently be considered under the following subheadings.

### 2.2.1  Sources

As we have seen, a **narrow line source** is required for AAS. Although in the early days vapour discharge lamps were used for some elements, these are rarely used now because they exhibit self-absorption. The most popular source is the hollow-cathode lamp, although electrodeless discharge lamps are popular for some elements.

#### 2.2.1.1  The hollow-cathode lamp

The hollow-cathode lamp is shown diagrammatically in Fig. 2.2. As the name suggests, the central feature is a **hollow cylindrical cathode**, lined

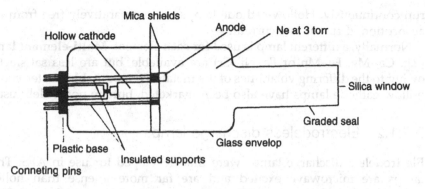

**Figure 2.2** The hollow-cathode lamp.

with the metal of interest. The lamp is contained within a glass envelope filled with an **inert gas** (usually Ne or Ar) at 1-5 Torr. A potential of about 500 V is applied between the electrodes and, at the pressures used, the discharge concentrates into the hollow cathode. Typically, currents of 2-30 mA are used. The filler gas becomes charged at the anode, and the ions produced are attracted to the cathode and accelerated by the field. The bombardment of these ions on the inner surface of the cathode causes metal atoms to **sputter** out of the cathode cup. Further collisions excite these metal atoms, and a simple, intense characteristic spectrum of the metal is produced. Marcus, and Kirkbright and Sargent (see Appendix C) describe this action and hollow-cathode lamps in more detail.

The **insulation** helps to confine the discharge within the hollow cathode, thus reducing the possibility of self-absorption and the appearance of ion lines. Both of these effects can cause bending of calibration curves towards the concentration axis. A glass envelope is preferred for ease of construction, but a silica window must be used for ultraviolet light transmission. A **graded seal** between the window and envelope ensures excellent gas tightness and shelf-life. A moulded plastic base is used. The choice of **filler gas** depends on whether the **emission lines** of the gas lie close to useful resonance lines and on the relative **ionization potentials** of the filler gas and cathode materials. The ionization potential of **neon** is higher than that of **argon**, and the neon spectrum is also less rich in lines. Therefore, neon is more commonly used.

Modern hollow-cathode lamps require only a very short **warm-up** period. **Lifetimes** are measured in ampere hours (usually they are in excess of 5 A h). A starting voltage of 500 V is useful, but operating **voltages** are in the range 150-300 V. In many instruments, the current supplied to the lamp is **modulated**. Hollow-cathode lamps may also be **pulsed** or run continuously. Hollow-cathode lamps are comparatively free from self-absorption, if run at low current.

Normally, a different lamp is used for each element. **Multi-element lamps** (e.g. Ca–Mg, Fe–Mn or Fe–Ni–Cr) are available, but are less satisfactory owing to the differing volatilities of the metals. **Demountable** (water-cooled) hollow-cathode lamps have also been marketed, but are not widely used.

### 2.2.1.2  Electrodeless discharge lamps

Electrodeless discharge lamps were first developed for use in AFS. These lamps are microwave excited and are far more intense than hollow-cathode lamps, but more difficult to operate with equivalent stability. **Radiofrequency-excited** electrodeless discharge lamps (the radiofrequency region extends from 100 kHz to 100 MHz, whereas the microwave region lies around 100 MHz) are typically less intense (only 5–100 times **more intense** than hollow-cathode lamps), but more reproducible. Commercially available radiofrequency lamps have a built-in starter (the starter provides a high-voltage spark to ionize some of the filler gas for initiation of the discharge), are run at **27 MHz** from a simple power supply (capable of supplying 0–39 W), pre-tuned and **enclosed** to stabilize the temperature and hence the signal.

A diagram of such a lamp is shown in Fig. 2.3 [taken from Barnett *et al.*, *At. Absorpt. Newsl.* **15**, 33 (1976). This paper gives a good account of the analytical performance of electrodeless discharge lamps].

High intensity is not a source requirement in AAS and therefore electrodeless discharge lamps will not replace hollow-cathode lamps. However, for those elements that produce poor hollow-cathode lamps (notably arsenic

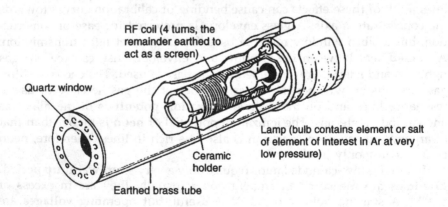

**Figure 2.3**  Cutaway diagram of an RF-excited electrodeless discharge lamp.

and selenium), the **signal-to-noise ratio**, because of the low signal, may so adversely affect detection limits that electrodeless discharge lamps can offer improvements. The analytical signal is the ratio of $I$ to $I_0$. Therefore, improved intensity of the signal can never improve sensitivity in AAS.

## 2.2.1.3 Source requirements in AAS

This leads us to a summary of source requirements in AAS. The source must give a **narrow resonance line profile** with little background, and should have a **stable** and **reproducible** output of sufficient intensity to ensure high **signal-to-noise ratios**. The source should be **easy to start**, have a **short warm-up** time and a long shelf-life.

---

**Q.** How are the metal atoms produced and excited in a hollow-cathode lamp?

---

**Q.** What is the normally preferred filler gas in a hollow-cathode lamp?

---

**Q.** Why must quartz windows be used in sources for AAS?

---

**Q.** What are the advantages of radiofrequency-excited electrodeless discharge lamps?

---

**Q.** Why does greater source intensity not lead to increased absorbance?

---

## 2.2.2 Flames

Several types of atom cell have been used for AAS. Of these, the most popular is still the flame, although a significant amount of analytical work is performed using various electrically heated graphite atomizers. This second type of atom cell is dealt with at length in Chapter 3, and the material here is confined to flames.

In AAS, the flame is only required to produce **ground-state atoms** (cf. AES, where a hot flame is preferred as atoms must also be excited). Frequently, an **air–acetylene** flame is sufficient to do this. For those elements which form more refractory compounds, or where interferences are encountered (see Section 2.4), a **nitrous oxide–acetylene** flame is preferred. In either case, a slot burner is used (100 mm for air–acetylene, 50 mm for nitrous oxide–acetylene) to increase the **path length** (this arises from Eqn. 2.3, Section 2.1) and to enable a specific portion of the flame to be viewed. Atoms are not uniformly distributed throughout the flame and, by adjusting the burner up and down with respect to the light beam, a **region of optimum absorbance** can be found.

This non-uniformity in the distribution of the atoms in the flame arises because the flame has a distinct structure. Figure 2.4 shows the structure of a typical **premixed flame**. Premixed gases are heated in the **preheating zone**, where their temperature is raised exponentially until it reaches the **ignition temperature**. Surrounding the preheating zone is the **primary reaction zone**, where the most energetic reactions take place.

The primary reaction zone is a hollow cone-like zone, only $10^{-5}$–$10^{-4}$ m thick. The actual **shape of the cone** is determined largely by the velocity distribution of the gas mixture leaving the burner. While the velocity of the gases at the burner walls is virtually zero, it reaches a maximum in the centre. The rounding at the top is caused, in part, by thermal expansion of the gases, which also produces a back-pressure which distorts the base

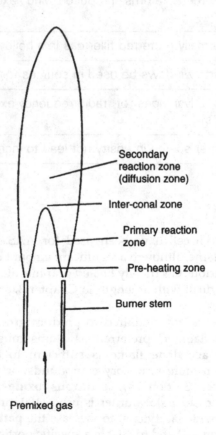

Secondary
reaction zone
(diffusion zone)

Inter-conal zone

Primary reaction
zone

Pre-heating zone

Burner stem

Premixed gas

**Figure 2.4** A premixed (or laminar) flame.

of the cone, causing some overhang of the burner. The cone elongates as the gas flow is increased. If this flow is increased so much that the gas velocity exceeds the burning velocity, the flame will **lift off**. If the flow is decreased so that the reverse happens, the flame will **strike back** with possibly explosive effect.

The primary reaction zone is so thin that thermodynamic equilibrium cannot possibly be established in it and the partially combusted gases and the **flame radicals** (e.g. OH*, H*, $C_2$* CH* and CN*), which propagate the flame pass into the **interconal zone**. Equilibrium is quickly established here as radicals combine. It is usually regarded as the hottest part of the flame and the most favoured for analytical spectrometry.

The hot, partly combusted gases then come into contact with oxygen from the air and the final flame products are formed. This occurs in what is known as the **secondary reaction zone** or diffusion zone. This discussion refers to premixed gas flames with **laminar** (i.e. non-turbulent) flow of the gas mixture to the flame.

---

**Q.** What are the requirements of a flame in AAS?

---

**Q.** Why are long slot burners preferred for AAS?

---

**Q.** Describe and explain the shape of the primary reaction zone.

---

**Q.** How can flames be prevented from striking back?

---

**Q.** Why does a kettle boil faster when the tips of the blue cones of the Bunsen burner flame are immediately below its base?

---

## 2.2.2.1 Flame temperatures

Various approaches to measuring flame temperature are well described in Gaydon's book on flames (see Appendix C). The best methods are spectroscopic rather than those which use thermocouples. The sodium **line reversal method** is perhaps the easiest. Sodium is added to the flame and the sodium D lines viewed against a bright continuum source (e.g. a hot carbon tube). When the flame is cooler than the source the lines appear dark because of absorption. When the flame is hotter than the tube, the bright lines stand out in emission. The current to the tube, which will have been precalibrated for temperature readings by viewing the tube with an optical pyrometer, is adjusted until the lines cannot be seen. At this reversal point, the flame and tube temperature should be equal.

Other methods, based upon **two lines**, may be used. Two-line methods may be used in absorption, emission or fluorescence. The signal is measured at the lines obtained when metal ions are sprayed into the flame. Provided there is no self-absorption and the transition probabilities of the lines are known accurately, the flame temperature can be calculated from the ratio of line intensities using the **Maxwell–Boltzmann distribution**.

Flame temperature varies from one part of the flame to another, as indicated in Section 2.2.2. Figures 2.5 and 2.6 show this effect for a stoichiometric and a fuel-rich flame, respectively.

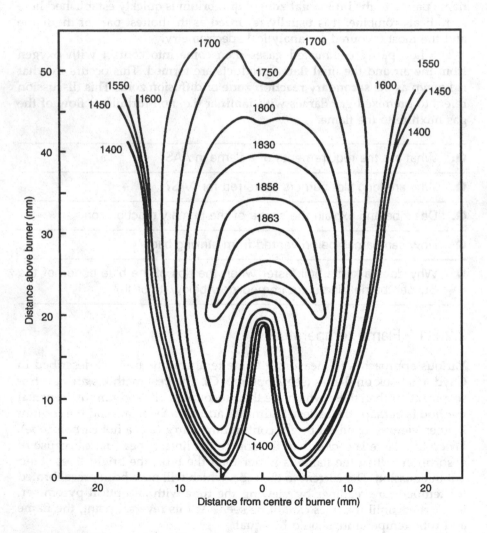

**Figure 2.5** Temperature distribution in a near stoichiometric premixed air–natural gas flame. Temperatures in °C.

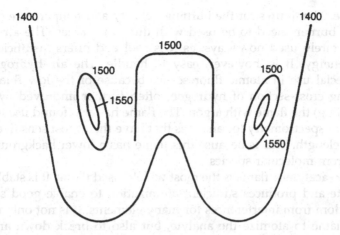

**Figure 2.6** Comparative figure (with Fig. 2.5) showing a fuel-rich flame temperature distribution. Temperatures in °C.

**Q.** List some ways in which flame temperature may be measured.

**Q.** Why does line reversal occur?

## 2.2.2.2 Flame gas mixtures

Table 2.1 lists some characteristics of the most popular premixed flames used for analytical spectrometry. These values should only be taken as

**Table 2.1** Characteristics of pre-mixed flames.

| | | Flow rates for stoichiometric flame (l min⁻¹) | | Maximum burning velocity | Approximate experimental temperature |
|---|---|---|---|---|---|
| Oxidant | Fuel | Oxidant | Fuel | (cm s⁻¹) | (K) |
| Air | Propane | 8 | 0.4 | 45 | 2200[a,b] |
| Air | Hydrogen | 8 | 6 | 320 | 2300[a,b] |
| Air | Acetylene | 8 | 1.4 | 160 | 2500[a,b,c] |
| Nitrous oxide | Acetylene | 10 | 4 | 285 | 3150[a,c] |

[a] A.G. Gaydon and H.G. Wolfhard, *Flames: Their Structure, Radiation and Temperature*, Chapman & Hall, London (1960), p. 304.
[b] J.B. Willis, *Appl. Opt.* **7**, 1295 (1968).
[c] K.M. Aldous, B.W. Bailey and J.M. Rankin, *Anal. Chem.* **44**, 191 (1972)

indicative. The figures for the burning velocity and temperature show that different burners need to be used with different flames. The **air–propane** flame is rarely used nowadays, as it is cool and offers insufficient atomization energy. It is, however, easy to handle. The **air–hydrogen** flame finds special use in atomic fluorescence because of the low fluorescence-quenching cross-section of hydrogen, often further improved by diluting (and cooling) the flame with argon. The flame has also found use for atomic absorption spectrometry for analytes that have their most sensitive line at a low wavelength. This is because this flame has a lower background signal arising from molecular species.

The **air–acetylene** flame is the most widely used flame. It is stable, simple to operate and produces sufficient atomization to enable good sensitivity and freedom from interferences for many elements. It is not only necessary for the flame to atomize the analyte, but also to break down any refractory compounds which may react with or physically entrap the analyte. Atomization, as we shall see, occurs both because of the high enthalpy and **temperature** of the flame, and through **chemical** effects. Thus, increasing the oxygen content of the flame above the approximately 20% normally present in air, while raising the flame temperature, does not necessarily enhance atomization, because more refractory oxides may be produced. Making the flame more fuel rich lowers the temperature but, by making the flame more reducing, increases the atomization of the elements such as molybdenum and aluminium.

The **nitrous oxide–acetylene** flame is both **hot** and **reducing**. A characteristic red, interconal zone is obtained under slightly fuel-rich conditions. This **red feather** is due to emission by the **cyanogen radical**. This radical is a very efficient scavenger for oxygen, thus pulling equilibria such as

$$TiO \rightleftharpoons Ti + \tfrac{1}{2}O_2$$

over to the right. This appears to be a vital addition to the high temperature which also promotes dissociation. Amongst those elements which are best determined in nitrous oxide–acetylene are Al, B, Ba, Be, Mo, Nb, Re, Sc, Si, Ta, Ti, V, W, Zr, the lanthanides and the actinides. The nitrous oxide–acetylene flame must be operated more carefully than the air–acetylene flame. For safety reasons, an air–acetylene flame is lit first, made very fuel rich and then switched to nitrous oxide by a two-way valve. Many modern instruments will perform this automatically. The flame is shut down by the reverse procedure. The nitrous oxide–acetylene flame can normally be run without any problems, provided that it is never run fuel lean and carbon deposits not allowed to build up. Any deposits should be cleaned away when the flame is extinguished.

**Q.** Why are different burners needed for different flames?

**Q.** What are the advantages and disadvantages of air–propane flames?

**Q.** Why is the air–acetylene flame so popular for AAS?

**Q.** What causes the red feather observed in the nitrous–oxide acetylene flame?

**Q.** What advantages does the nitrous oxide–acetylene flame offer for AAS?

## 2.2.3 Sample introduction and sample atomization

So far we have no analyte atoms in the atom cell! This is usually achieved in the following manner, although some alternative ways are considered in Chapter 7.

Figure 2.7 shows a typical **pneumatic nebulization** system for a premixed flame. The sample is sucked up a plastic **capillary tube**. In the type of **concentric nebulizer** illustrated here, the sample liquid is surrounded by the oxidant gas as it emerges from the capillary. The high velocity of this gas, as it issues from the tiny annular orifice, creates a pressure drop which sucks up, draws out and 'shatters' the liquid into very tiny droplets. This phenomenon is known as the **venturi effect** and is illustrated in Fig. 2.8.

The nebulizer capillary position may be adjustable on a screw thread to permit optimization of sample uptake and drop size. Alternatively or additionally, an **impact bead** may be placed in the path of the initial aerosol to provide a secondary fragmentation and so improve the efficiency of nebulization. Such a device is illustrated in Fig. 2.9.

The material of the nebulizer must be highly **corrosion resistant**. Commonly, the plastic capillary is fixed to a platinum–iridium alloy (90:10) capillary mounted in stainless-steel gas supply inlets. The impact bead is sometime made of a similar alloy or smooth borosilicate glass.

The aerosol then passes along the plastic expansion chamber. Large droplets collect on the walls of the chamber and, to ensure that only the smallest particles reach the flame, **spoilers** or **baffles** may be placed in the path of the gases. The chamber also allows for mixing of the gases and tends to damp fluctuations in nebulization efficiency. Some loss of solvent by evaporation will also occur. The chamber requires a drain tube which must be sealed to provide a back-pressure for the flame. This is usually

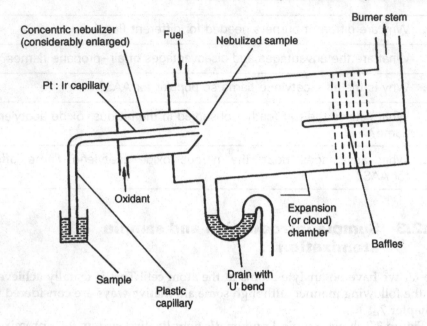

**Figure 2.7** Schematic diagram of a concentric nebulizer system for a premixed burner.

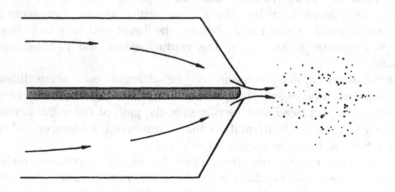

**Figure 2.8** The venturi effect.

achieved by a hydraulic seal (a 'U'-tube). It is imperative that water is present in the 'U'-tube before lighting the flame!

The **efficiency** of this kind of indirect nebulizer and expansion chamber arrangement is usually about 10–15%. It is favoured because only very small droplets reach the flame, but clearly it would be useful to improve its

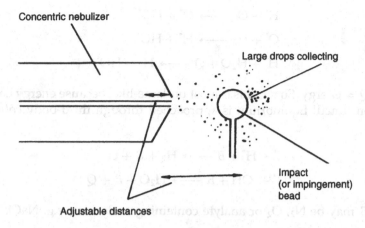

**Figure 2.9**  Impact bead assembly.

efficiency. The use of **organic solvents** is known to do this. Organic solvents may affect the nebulization because their different surface tension, density, viscosity or vapour pressure, all of which enter into the relevant equations, probably results in smaller droplets which evaporate more quickly. They also affect results because of their effect on the flame (they essentially act as a secondary fuel).

At this stage no atoms have been formed — only a mist. Hence the name nebulization, not atomization. We will now discuss **atomization**.

The **desolvation** of the droplets is usually completed in the preheating zone. The mist of salt clotlets then fuses and **evaporates** or sublimes. This is critically dependent on the size and number of the particles, their composition and the flame mixture. As the absolute concentration of analyte in the flame is very small ($< 10^{-3}$ atm), the saturated vapour pressure may not be exceeded even at temperatures below the melting point.

Many of these vapours will **break down** spontaneously to atoms in the flame. Others, particularly diatomic species such as metal **monoxides** (e.g. alkaline earth and rare earth oxides), are more refractory. Monohydroxides which can form in the flame can also give problems. The high temperature and enthalpy of the flame aid **dissociation** thermodynamically, as does a reducing environment. The role **of flame chemistry** is also important. Atoms, both ground state and excited, may be produced by radical reactions in the primary reaction zone. If we take the simplest flame (a hydrogen–oxygen flame), some possible reactions are the following:

$$H_2 + Q \longrightarrow 2H^{\bullet}$$

$$O_2 + Q \longrightarrow 2O^{\bullet}$$

$$H^\bullet + O_2 \longrightarrow O^\bullet + HO^\bullet$$

$$O^\bullet + H_2 \longrightarrow H^\bullet + HO^\bullet$$

$$H^\bullet + H_2O + Q \longrightarrow H_2 + HO^\bullet$$

where $Q$ = energy. So far, this is most unflame-like, because energy has only been consumed! Equilibrium is approached through third-body collisions, such as

$$H^\bullet + H^\bullet + B \longrightarrow H_2 + B + Q$$

$$H^\bullet + {}^\bullet OH + B \longrightarrow H_2O + B + Q$$

where $B$ may be $N_2$, $O_2$ or analyte containing molecules, e.g. NaCl:

$$H^\bullet + NaCl \longrightarrow Na^\bullet + HCl$$

$$H^\bullet + {}^\bullet OH + NaCl \longrightarrow H_2O + Na^\bullet + Cl^\bullet + Q$$

$$H^\bullet + {}^\bullet OH + NaCl \longrightarrow H_2O + Na^{\bullet *} + Cl^\bullet$$

where the asterisk denotes an atom excited by 'chemiluminescence'.

For many elements, the **atomization efficiency** (the ratio of the number of atoms to the total number of analyte species, atoms, ions and molecules in the flame) is 1, for others it is less than 1, even for the nitrous oxide–acetylene flame (for example, it is very low for the lanthanides). Even when atoms have been formed they may be lost by compound formation and **ionization**. The latter is a particular problem for elements on the left of the Periodic Table (e.g. $Na \rightarrow Na^+ + e^-$: the ion has a noble gas configuration, is difficult to excite and so is lost analytically). Ionization increases exponentially with increase in **temperature**, such that it must be considered a problem for the alkali, alkaline earth, and rare earth elements and also some others (e.g. Al, Ga, In, Sc, Ti, Tl) in the nitrous oxide–acetylene flame. Thus, we observe **some self-suppression** of ionization at higher concentrations. For trace analysis, an **ionization suppressor** or **buffer** consisting of a large excess of an easily ionizable element (e.g. caesium or potassium) is added. The excess caesium ionizes in the flame, suppressing ionization (e.g. of sodium) by a simple, mass action effect:

$$Cs \longrightarrow Cs^+ + e^-$$

$$Na \longleftarrow Na^+ + e^-$$

Differing amounts of easily ionizable elements in real samples cause varying ionization suppression and hence the possibility of interference (see Section 2.4.2).

**Q.** How does the venturi effect contribute to the nebulization of the sample?

**Q.** In the type of nebulizer described, where does most of the sample go?

**Q.** How can ionization problems be overcome?

**Q.** Summarize nebulization and atomization in a schematic diagram. Now compare this with Fig. 2.10.

## 2.2.4  Burner design

Many early experiments were carried out with the **Meker type** of burner. A perforated metal plate is placed on top of an open pipe (the thicker the plate, the more stable is the flame). On ignition, a number of small cones

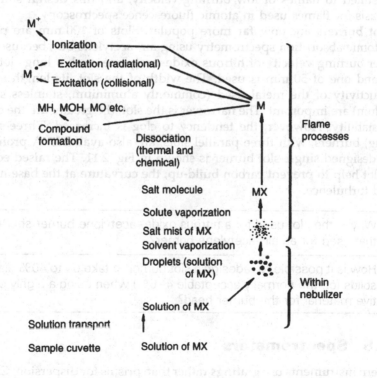

**Figure 2.10**  Summary of atomization in flames.

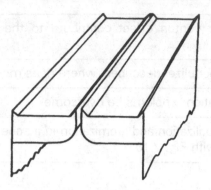

**Figure 2.11** Cutaway diagram to show the profile of a single-slot burner with raised edges.

are formed over each opening. During design, the number of holes, their size and conductivity of the metal must be considered. Such burners are best suited to flames of low burning velocity, and this design still forms the basis for flames used in atomic fluorescence spectroscopy.

**Slot burners** are now far more popular. Slots of 100 mm are popular for atomic absorption spectrometry using air–acetylene but, because of the higher burning velocity of nitrous oxide–acetylene, such a long slot is not safe and one of 50 mm is used. The width of the slot, its length and the conductivity of the metal used (commonly aluminium, stainless steel or titanium) are important. The narrower is the slot, the greater are the cooling and stability. However, the tendency to clog is increased. **Three-slot** (or Boling) burners, with three parallel slots, are also available. A profile of a well designed single-slot burner is shown in Fig. 2.11. The raised edges at the slot help to prevent carbon build-up; the curvature at the base helps to avoid turbulence.

---

**Q.** Why is the slot used for a nitrous oxide–acetylene burner shorter than that used for an air–acetylene burner?

---

**Q.** How is it possible to redesign a slot burner to take up to 40% dissolved solids (cf. the normally acceptable 4–6%) when using a highly conductive material for the burner head?

---

## 2.2.5 Spectrometers

Modern instruments use **gratings** rather than prisms for dispersion; Czerny–Turner and Ebert systems are commonly employed (see Section 4.4.5).

As only **moderate resolution** is needed (because of the 'lock and key' effect), focal lengths of 0.25–0.50 m and rulings of 600–3000 lines mm$^{-1}$ are commonly employed. Resolution in the region 0.2–0.02 nm are claimed.

As atomic emission and atomic absorption take place at the **same wavelength**, it is useful to be able to discriminate between the two signals to maximize atomic absorption sensitivity. Figure 2.12 shows how this may be done. In Walsh's first design (Fig. 2.12a), the light from the lamp passes through the flame (the source being at the focus of the first lens) in a parallel beam which is focused on the entrance slit of the monochromator, this being at the focus of the second lens. The flame is placed at the focus of the second lens so that flame emission is exactly defocused at the monochromator. Thus, the atomic absorption signal is maximized and the atomic emission signal minimized.

The next development was **modulation**. A **rotating sector** (often crudely referred to as a chopper) is placed in the light beam (Fig. 2.12b). As the

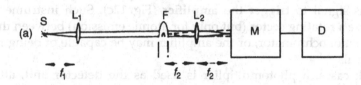

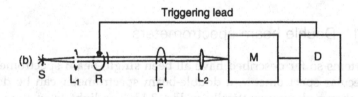

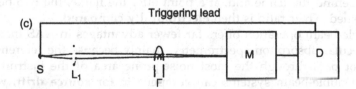

**Figure 2.12** Optical and instrumental arrangements for AAS. S = source; L$_1$ = first lens; L$_2$ = second lens; F = flame; $f$ = focal length; R = rotating sector; M = monochromator; D = detector.

beam strikes the solid part of the chopper, it is interrupted; the hole in the centre allows it to pass. The sector is placed between the source and the flame. The atomic absorption signal is now modulated, but the atomic emission signal is not. An **AC amplifier** tuned to the atomic absorption signal, via **phasing coils** on the rotating sector, **gates** the amplifier and selectively amplifies the atomic absorption signal as opposed to the DC atomic emission signal. It is therefore essential that the sector be placed between the source and the flame. In atomic absorption and atomic emission spectrometers, two rotating sectors may be found, one for atomic absorption and one for atomic emission (between the flame and the monochromator). The atomic emission chopper only functions when the source is off and is necessary if the instrument is fitted with an AC amplifier. When the instrument is switched back to atomic absorption mode, the sector should lodge in the open mode.

Often flame radiation may be reflected from the back of a rotating sector, which is less than totally reliable because it is mechanical. Most modern instruments therefore **modulate the power** applied to the source and also use this signal to **trigger the amplifier** (Fig. 12c). Such instruments may still have a rotating sector (but only for atomic emission) between the flame and the monochromator, or the amplifier may be capable of being reset for DC signals.

In all cases, a photomultiplier is used as the detector and, after suitable amplification, a variety of read-out devices may be employed (see Section 2.2.5.3).

### 2.2.5.1  Double beam spectrometers

The systems so far described have all been **single-beam** spectrometers. As in molecular spectrometry, a double-beam spectrometer can be designed. This is shown diagrammatically in Fig. 2.13. The light from the source is split into **two beams**, usually by means of a rotating **half-silvered mirror** or by a **beam splitter** (a 50%-transmitting mirror). The second reference beam passes behind the flame and, at a point after the flame, the two beams are recombined. Their ratio is then **electronically compared**.

Double-beam operation offers far fewer advantages in AAS than it does in molecular absorption spectrometry, mainly because the reference beam does not pass through the most noise-prone area of the instrument, the flame. Double-beam systems can compensate for **source drift**, warm-up and **source noise**. This should lead to improved **precision** and often does. However, as the major source of noise is likely to be the flame, this advantage is slight and may be more than offset by the significant loss of intensity in the light signal, and hence lower signal-to-noise ratio.

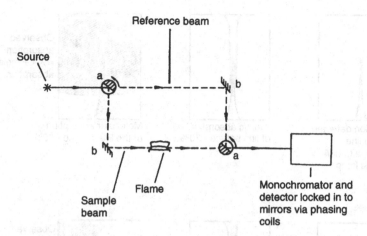

**Figure 2.13** Double-beam atomic absorption instrumentation: a = rotating half-silvered mirror; b = front surface mirror.

## 2.2.5.2 Background correction

Considerably more advantage can be derived from the use of a second beam of **continuum radiation** to correct for non-atomic absorption. Figure 2.14 schematically summarizes how this operates. When using a **line source** such as a hollow-cathode lamp, we observe **atomic absorption** in the flame, **absorption from molecular species** and scattering from particulates. The latter, known as **non-specific absorption**, is a particular problem at shorter wavelengths and can lead to positive errors. When using a **continuum source** (e.g. a deuterium arc or hydrogen hollow-cathode lamp), the amount of atomic absorption observed, as we have already seen (Section 2.1), is negligible, but the **same amount of non-specific absorption** is seen. Thus, if the signal observed with the continuum source is **subtracted** from that observed with the line source, the error is removed.

Figure 2.15 shows an instrument capable of doing this **simultaneously and automatically**. Lead is particularly prone to this problem, and Fig. 2.16 shows how background correction can be used to remove the interference of non-specific absorption when determining lead in chromium. Notice that the **precision** is also improved, mainly because the effects which give rise to the background are not very reproducible.

Other types of background correction have also been developed. The **Zeeman** effect background correction system started gaining popularity in the early 1980s. An atomic spectral line when generated in the presence of a **strong magnetic field** can be split into a number of components

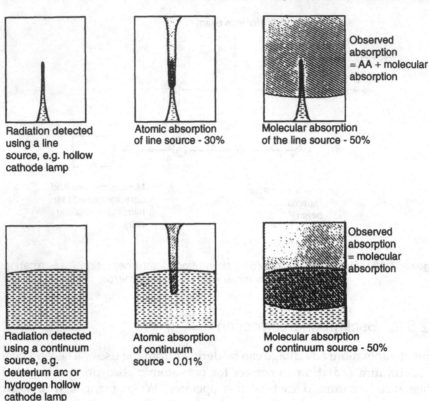

Radiation detected using a line source, e.g. hollow cathode lamp

Atomic absorption of line source - 30%

Molecular absorption of the line source - 50%

Observed absorption = AA + molecular absorption

Radiation detected using a continuum source, e.g. deuterium arc or hydrogen hollow cathode lamp

Atomic absorption of continuum source - 0.01%

Molecular absorption of continuum source - 50%

Observed absorption = molecular absorption

Corrected atomic absorption = Absorption observed using line source - absorption observed using continuum source

**Figure 2.14** Background correction.

of slightly different wavelength. The field can be applied in a number of different ways. It may be applied to either the source or the atom cell and in either a **transverse** (magnetic field and measurement beam are 90° to each other) or a **longitudinal** (magnetic field and measurement beam are parallel) configuration. In the simplest form of transverse Zeeman effect, the line appears as **three components** (Fig. 2.17). The $\pi$ **component** is situated at the 'normal' wavelength of the line, but the $\sigma^+$ and $\sigma^-$ **components** lie an equal distance on either side. The $\sigma$ components are **linearly polarized** perpendicular to the magnetic field. If the field is strong enough, these components will lie outside the atomic absorption profile and background can be corrected by measuring the absorbance of the $\pi$ and $\sigma$ components, respectively. In the longitudinal configuration, only the two shifted lines

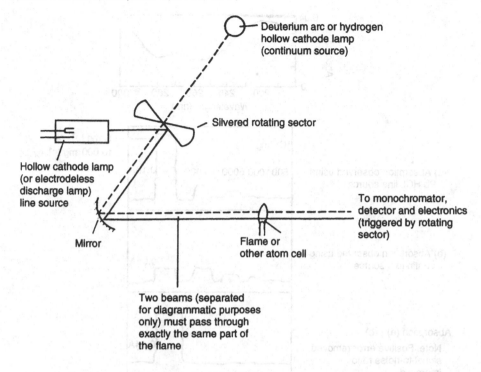

**Figure 2.15** Automatic simultaneous background corrector.

are seen. The non-shifted $\pi$ component is not seen. The magnet operates at about 50–60 Hz, and each measurement cycle is comprised of two phases. The magnet is off for one of them and on for the other. When the magnet is off, the atomic absorption is superimposed on molecular and non-specific absorption. This therefore contains both atomic and background absorption signals. When the magnet is on, the lines are split and the two $\sigma$ components have no atomic absorption but do have background absorption. The atomic signal may therefore be obtained by subtraction of this signal from the total signal (i.e. signal with magnet off − signal with magnet on). The strength of the magnetic field of Zeeman spectrometers is typically 0.9–1.0 T, which shifts the lines approximately 0.01 nm either side of the original wavelength.

Another type of background correction system that has found some use is that developed by Smith and Hieftje. The **Smith–Hieftje** background correction technique is of especial use when there is strong molecular interference, such as that observed by phosphate on selenium or arsenic determinations. If the hollow-cathode lamp is run at its normal operating

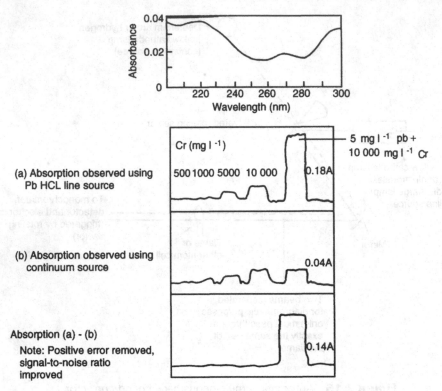

**Figure 2.16** Background correction in action. (Top) background due to spraying
chromium blank. (Bottom) determination of Pb in Cr.

current, then atomic and molecular absorption by the analyte and
interferents in the atom cell will occur. If, however, the lamp is **pulsed**
periodically to much higher currents, self-absorption in the lamp prevents
virtually all atomic absorption in the atom cell, but leaves the molecular
absorption. The corrected atomic absorption signal may therefore be
obtained by subtraction. This method of background correction, although
popular and successful for a while, does lead to accelerated wear of the
hollow cathode lamps.

**Q.** How can the interference of AA measurements by AE signals be
removed?

**Q.** What are the advantages in AAS of (i) modulation, (ii) double-beam
systems and (iii) background correction?

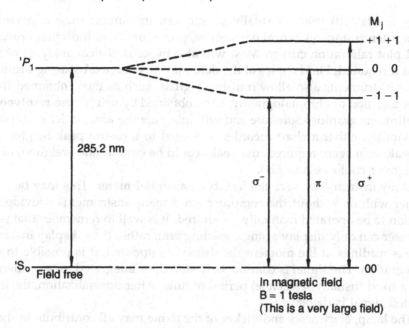

285.2 nm

Field free

In magnetic field
B = 1 tesla
(This is a very large field)

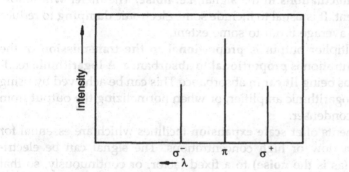

(The diagram exaggerates the separation of components
for clarity. In this instance, it would be 0.0038 nm.)

**Figure 2.17** Normal Zeeman effect for magnesium.

## 2.2.5.3 Read out systems

After suitable amplification, the signal may be read out using a variety of
different approaches. Older instruments used an **analogue meter** as a means
of read-out. Considerable advances have been made in read-out systems

over the last 10 years. Virtually all modern instruments use a **personal computer** to store analytical parameters, procedures and individual results, and plot calibration curves. Most will also make statistical analyses of the data if required. Hard copies of the data may be obtained by using printers. Most instruments also allow transient signals, such as those obtained from flow injection or chromatography, to be obtained by using **time-resolved** or **continuous graphics** software and will integrate the area under each peak. Previously, either a **chart recorder** was used to measure peak heights (or, if peak areas were required, the peaks could be cut out and weighed) or an integrator could be attached.

Many instruments have the facility of a **digital meter**. This may be used either with or without the computer since many instruments provide the option to be operated manually if required. It is well to remember that such a meter can only display a single reading and, rather than display instantaneous readings at the moment the display is updated, it is possible to use **integration**. The signal is continuously stored, or integrated, in a condenser for a fixed (usually selectable) period of time. After normalization, the integrated signal is displayed.

The lamp, electronics and **flicker** of the flame may all contribute to short-term, **irregular fluctuations** in the signal, i.e. **noise**. The meter will follow this to some extent. It is usual to include some electronic **damping** to reduce the noise, i.e. to average it out to some extent.

The photomultiplier output is **proportional to the transmission** of the flame, yet concentration is **proportional to absorbance**. A **logarithmic** read-out is preferred as being **linear** in absorbance This can be achieved by using an appropriate logarithmic amplifier, or when normalizing the output from the integrating condenser.

Many instruments offer **scale expansion** facilities which are essential for **accurate work** at low or high concentrations. The signal can be electrically expanded (as is the **noise**) to a fixed factor, or continuously, so that the reading of a chosen standard comes to a desired figure (e.g. 10 μg ml$^{-1}$ reads 100). This latter option is referred to as '**concentration read-out**'. Care must be taken when using this at high concentrations, as few calibration curves are still linear above 0.5 absorbance. Curves tend to **bend towards** the concentration axis, principally because **stray light** (i.e. unabsorbable light, for example from nearby lines in the lamp) become an important contribution to the total light falling on the detector. Controls for '**curve correction**' may be supplied which enable the calibration to be **linearized** by extra expansion on standards of high concentration. Many modern instruments provide an autosampler, and if this is in use then, for samples whose concentration exceeds the linear part of the calibration curve, the analytical procedure may be programmed so that automatic dilution is performed.

The computer will then multiply the concentration obtained by the relevant dilution factor automatically, and the concentration in the undiluted sample will be read out.

In many instruments, the computer will control the instrument settings (if a coded hollow-cathode lamp or a procedure that has been stored in the memory is being used, the computer may set the wavelength automatically). In addition, the recommended flame composition and the height at which readings will be taken may also be optimized.

---

**Q.** In AAS, where does noise come from?

---

**Q.** How can noise problems be minimized?

---

**Q.** What is the advantage of logarithmic read-out in AAS?

---

**Q.** When is curve correction needed and how can it be performed?

---

## 2.3   SENSITIVITY AND LIMIT OF DETECTION

**Limits of detection** for FAAS are typically in the range 0.01–0.1 $\mu g\ ml^{-1}$. A full list is given in Table 2.2. The linear dynamic range (i.e. the maximum range over which the calibration curve is linear) is limited by the propensity for self-absorption in the flame, and is generally no more than three orders of magnitude (e.g. from 0.01 to 10 $\mu g\ ml^{-1}$)

## 2.4   INTERFERENCES AND ERRORS

The use of a **line source** and the **ratio method** (i.e. $I_0/I$) tend to minimize errors in AAS. Thus, if the **wavelength setting** is seriously incorrect, it is unlikely that any absorption will be observed. If the wavelength is incorrectly tuned, the effects on the value of $I$ will roughly equal those on the value of $I_0$, and the error may not be too serious.

Since AAS is a **ratio method**, many instrumental errors (e.g. long-term source drift, small monochromator drifts) should cancel out, as $I$ is ratioed to $I_0$. However, a stable uptake rate, or **aspiration rate**, is required. This falls as the viscosity of the solution sprayed is increased. Nebulizer uptake interferences can be minimized if the **dissolved salts** content of samples and standards is approximately matched. For example, when determining $\mu g\ cm^{-3}$ sodium levels in 2 M phosphoric acid, ensure that the standards are also dissolved in 2 M phosphoric acid, using a blank to check for contamination.

**Table 2.2** Limits of detection for flame atomic absorption spectrometry.

| Element | | Characteristic concentration† ($\mu g \ ml^{-1}$) | Detection limit ($\mu g \ ml^{-1}$) |
|---|---|---|---|
| Ag | Silver | 0.029 | 0.002 |
| Al | Aluminium | 0.75 | 0.018 |
| As | Arsenic | 0.92 (a) | 0.11 (a) |
| | | 0.6 (b) | 0.26 (b) |
| | | | 0.002 (c) |
| Au | Gold | 0.11 | 0.009 |
| B | Boron | 8.4 | 2.0 |
| Ba | Barium | 0.20 | 0.020 |
| Be | Beryllium | 0.016 | 0.0007 |
| Bi | Bismuth | 0.20 | 0.046 |
| Ca | Calcium | 0.013 | 0.002 |
| Cd | Cadmium | 0.011 | 0.0007 |
| Co | Cobalt | 0.053 | 0.007 |
| Cr | Chromium | 0.055 | 0.005 |
| Cs | Caesium | 0.040 | 0.004 (a) |
| Cu | Copper | 0.040 | 0.002 |
| Dy | Dysprosium | 0.67 | 0.028 |
| Er | Erbium | 0.46 | 0.026 |
| Eu | Europium | 0.34 | 0.014 |
| Fe | Iron | 0.045 | 0.006 |
| Ga | Gallium | 0.72 (a) | 0.038 (d) |
| Gd | Gadolinium | 19 | 1.1 |
| Ge | Germanium | 1.3 | 0.11 |
| Hf | Hafnium | 10 | 1.4 |
| Hg | Mercury | 2.2 | 0.16 |
| | | 0.0001 (c) | 0.00004 (c) |
| Ho | Holmium | 0.76 | 0.035 |
| In | Indium | 0.17 | 0.038 |
| Ir | Iridium | 2.3 | 0.36 |
| K | Potassium | 0.009 | 0.002 |
| La | Lanthanum | 48 | 2.1 |
| Li | Lithium | 0.017 | 0.0015 |
| Lu | Lutetium | 7.9 | 0.30 |
| Mg | Magnesium | 0.003 | 0.0002 |
| Mn | Manganese | 0.021 | 0.002 |
| Mo | Molybdenum | 0.28 | 0.030 |
| Na | Sodium | 0.003 | 0.0002 |
| Nb | Niobium | 19 | 2.9 |
| Nd | Neodymium | 6.3 | 1.1 |
| Ni | Nickel | 0.050 | 0.008 |
| Os | Osmium | 1.2 | 0.12 |
| Pb | Lead | 0.11 | 0.015 |
| Pd | Palladium | 0.092 | 0.016 |
| Pr | Praseodymium | 18 | 8.3 |

**Table 2.2** (continued).

| Element | | Characteristic concentration† ($\mu g \ ml^{-1}$) | Detection limit ($\mu g \ ml^{-1}$) |
|---|---|---|---|
| Pt | Platinum | 1.2 | 0.090 |
| Rb | Rubidium | 0.030 | 0.002 |
| Re | Rhenium | 9.5 | 0.85 |
| Rh | Rhodium | 0.12 | 0.005 |
| Ru | Ruthenium | 0.72 | 0.087 |
| Sb | Antimony | 0.29 | 0.041 |
| Sc | Scandium | 0.27 | 0.025 |
| Se | Selenium | 0.38 (a) | 0.25 (a) |
| | | | 0.04 (e) |
| | | | 0.002 (c) |
| Si | Silicon | 1.5 | 0.20 |
| Sm | Samarium | 6.6 | 0.75 |
| Sn | Tin | 0.39 | 0.031 |
| Sr | Strontium | 0.041 | 0.002 |
| Ta | Tantalum | 11 | 1.8 |
| Tb | Terbium | 7.9 | 0.5 |
| Te | Tellurium | 0.26 | 0.035 |
| Ti | Titanium | 1.4 | 0.05 |
| Tl | Thallium | 0.28 | 0.013 |
| Tm | Thulium | 0.27 | 0.014 |
| U | Uranium | 113 | 59 |
| V | Vanadium | 0.75 | 0.05 |
| W | Tungsten | 5.8 | 0.52 |
| Y | Yttrium | 2.3 | 0.11 |
| Yb | Ytterbium | 0.073 | 0.0021 |
| Zn | Zinc | 0.009 | 0.001 |
| Zr | Zirconium | 9.1 | 1.0 |

† Characteristic concentrations and detection limits quoted were all measured at the most sensitive line. Hollow cathode lamps were used throughout together with an R446 photomultiplier. Aqueous solutions were employed for all elements and the fuel/support gas combinations were those recommended for general analytical use, except for the special cases noted.

(a) Air-acetylene; (b) Air-hydrogen; (c) Vapour generation; (d) Nitrous oxide-acetylene; (e) Nitrogen-hydrogen-entrained air

**Q.** Why are errors not really associated with monochromator settings in AAS?

**Q.** Why is the observed absorbance of 2 $\mu g \ ml^{-1}$ zinc dissolved in dilute hydrochloric acid greater than that of 2 $\mu g \ ml^{-1}$ zinc dissolved in 1 M potassium dichromate solution?

## 2.4.1 Spectral interferences

These are the only type of interference that do not require the presence of analyte. For AAS the problem of spectral interference is not very severe, and **line overlap interferences are negligible**. This is because the resolution is provided by the 'lock and key' effect. To give spectral interference the lines must not merely be within the bandpass of the monochromator, but actually overlap each other's **spectral profile** (i.e. be within 0.01 nm). West [*Analyst* **99**, 886, (1974)] has reviewed all the reported (and a number of other) 'spectral interferences' in AAS. Most of them concern lines which would never be used for a real analysis, and his conclusion is that the only 'real' problem is in the analysis of copper heavily contaminated with europium! The most commonly used copper resonance line is 324.754 nm (characteristic concentration 0.1 $\mu g\ cm^{-3}$) and this is overlapped by the europium 324.753 nm line (characteristic concentration 75 $\mu g\ cm^{-3}$).

Spectral interferences from the overlap of **molecular bands and lines** (e.g. the calcium hydroxide absorption band on barium at 553.55 nm) cannot be so easily dismissed. Lead seems to be particularly prone to such **non-specific absorption** problems at the 217.0 nm line (e.g. sodium chloride appears to give strong molecular absorption at this wavelength). This type of problem is encountered in practical situations, but can sometimes be removed by the technique of **background correction** (see Section 2.2.5.2).

**Q.** Why can it be said that spectral interferences can be virtually eliminated in AAS?

**Q.** How can non-specific absorption problems in AAS be overcome?

## 2.4.2 Ionization interferences

Also called **vapour-phase interferences** or **cation enhancement**. In the air–acetylene flame, the intensity of rubidium absorption can be doubled by the addition of potassium. This is caused by **ionization suppression** (see Section 2.2.3), but if uncorrected will lead to substantial positive errors when the samples contain easily ionized elements and the standards do not. An example is when river water containing varying levels of sodium is to be analysed for a lithium tracer, and the standards, containing pure lithium chloride solutions, do not contain any ionization suppressor.

The problem is easily overcome by adding an **ionization suppressor** (or **buffer**) in large amount to all samples and standards.

**Q.** Why is ionization interference severe in AAS?

**Q.** What precautions with regard to both standards and samples would be needed when determining potassium in sea-water?

## 2.4.3 Chemical interferences

Given how easily the two types of interference discussed above can be overcome, this third type constitutes the biggest source of problems in AAS. A brief discussion is given of **solid-phase interferences** centred around the following **classification**:

(a) depressions caused by the formation of less volatile compounds which are difficult to dissociate;
(b) enhancements caused by the formation of more volatile compounds;
(c) depressions due to occlusion into refractory compounds;
(d) enhancements due to occlusion into more volatile compounds.

### (a) Formation of less volatile compounds

The best known of this type of interference is that of **phosphate on calcium**; sulphate and silicate have the same effect. Figure 2.18 illustrates the effect of increasing phosphate concentration on the calcium signal. The graph shows a pronounced 'knee' at ratios of P:Ca variously reported in the range 0.3–1.1. The effect is less pronounced higher up the flame and is not observed in a hotter flame, such as the nitrous oxide–acetylene flame. The effects are seen at the 422.7 or 620.0 nm line or the calcium hydroxide band at 554 nm. The **constant level** of interference observed above a certain level strongly suggests formation of a compound (probably a calcium phosphate). This compound is less volatile than calcium chloride, and hence the formation of calcium atoms is hindered.

There are many examples of this type of interference, several involving **aluminates**. They all show the pronounced 'knee' in the graph — this distinguishes them from non-specific occlusions.

Several **approaches** can be made to **reduce** such interferences:

(i) use a **hotter flame**;
(ii) **adjust the nebulizer** to produce a smaller particle size;
(iii) make **observations higher** in the flame;
(iv) use a **'releasing agent'**, an element that will enter into a 'law of mass action competition' with the analyte to combine with the interferent;

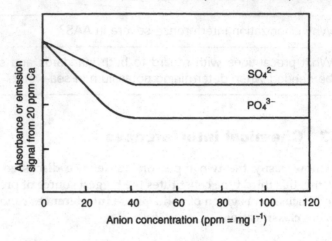

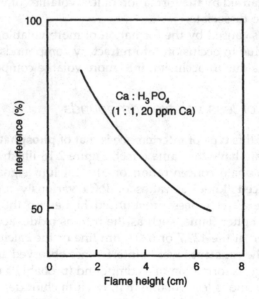

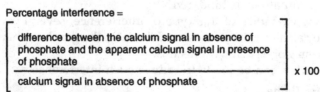

Percentage interference =

$$\left[ \frac{\text{difference between the calcium signal in absence of phosphate and the apparent calcium signal in presence of phosphate}}{\text{calcium signal in absence of phosphate}} \right] \times 100$$

**Figure 2.18** Anion interferences on calcium.

if an excess of the releasing agent is added, the analyte is released from the interfering anion (e.g. excess lanthanum or strontium releases calcium from phosphate interference);

(v)  use a **protective chelating agent** which preferentially complexes the analyte, protecting it from the 'grasp' of the interferent (e.g. excess EDTA protects calcium from phosphate interference).

## (b)  Formation of more volatile compounds

These interference effects are far less common. Under this heading, some authors classify the **enhancement** of signals from several, otherwise refractory, elements by fluoride. The use of protective agents (e.g. EDTA for calcium or 8-hydroxyquinoline for aluminium or chromium) are also examples of this type of effect.

## (c)  Occlusion into refractory compounds

Such depressions can be encountered when the matrix is refractory (e.g. zirconium, uranium or a rare earth element), and the small amount of analyte can be physically trapped in clotlets of matrix oxide in the flame. Such systems do not show a 'knee' [see type (a)] and can be minimized by higher flame temperature.

## (d)  Occlusion into volatile compounds

Some compounds (e.g. ammonium chloride) explosively **sublime** in the flame, thus enhancing atomization. By adding excess ammonium chloride to all samples and standards, this effect can be used to minimize interferences of types (a) and (c).

---

**Q.**  How can specific and non-specific depressions [types (a) and (c)] be distinguished?

---

**Q.**  List several ways in which the interference of phosphate on calcium may be minimized?

---

**Q.**  How might aluminium interfere with magnesium during AAS?

---

**Q.**  Iron depresses atomic absorption by chromium in the air–acetylene flame. For what reasons do workers add 8-hydroxyquinoline and/or ammonium chloride to minimize this interference?

---

## 2.4.4  Applications

Flame atomic absorption spectrometry can be used to determine trace levels of analyte in a wide range of sample types, with the proviso that the sample is first brought into solution. The methods described in Section 1.6 are all applicable to FAAS. Chemical interferences and ionization suppression cause the greatest problems, and steps must be taken to reduce these (e.g. the analysis of sea-water, refractory geological samples or metals). The analysis of oils and organic solvents is relatively easy since these samples actually provide fuel for the flame; however, build-up of carbon in the burner slot must be avoided. Most biological samples can be analysed with ease provided that an appropriate digestion method is used which avoids analyte losses.

# 3 ELECTROTHERMAL ATOMIZATION

## 3.1 HISTORICAL DEVELOPMENT

In 1905 and 1908, **King**, generally regarded as the first worker in this field, reported on fundamental spectral studies using an electrically heated tubular furnace. The classical work in analytical chemistry is that of **L'Vov**, who began to publish his results in 1959. Here the sample was applied to the tip of a **carbon electrode** which was introduced into a cylindrical **heated furnace** through a transverse aperture at the centre of the tube. At first, the sample was preheated using a powerful DC arc (arced to the electrode with the sample in position). Later this arrangement was replaced by simpler **resistive heating of the electrode** (Fig. 3.1). The graphite tube was 30–50 mm in length, with an internal diameter of 2.5–5.0 mm and an external diameter of 6.0 mm. At first, the cylinders were lined with tungsten or tantalum foil to retard vapour diffusion, but later this was changed to a coating of **pyrolytic graphite**. The tube did not act as an atomizing furnace, merely as an **atom cell** hindering the loss of atoms. The sample electrode was responsible for atomization. Both the tube and the electrode were heated via **step-down transformers** of 4 kW (at 10 V) and 1 kW (at 15 V), respectively. The atomizer was placed within a chamber filled with **argon** or **nitrogen** and light passed down the tube for AAS measurements. L'Vov's apparatus offers some of the best absolute sensitivities for AAS yet obtained, but has been criticized for being too cumbersome. Current commercial atomizers are based on a simpler design.

In 1967, **Massmann** described a **heated graphite furnace** in which no auxiliary electrode was used, i.e. the graphite tube was both the resistance element and the furnace. The sample was micro-pipetted directly into a 55 mm long, 6.5 mm internal diameter, 1.5 mm wall thickness tube via a small 2 mm diameter orifice. The absorption tube device and a graphite

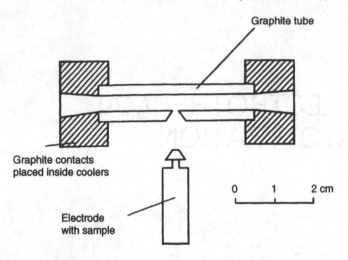

**Figure 3.1** Graphite furnace design of L'Vov.

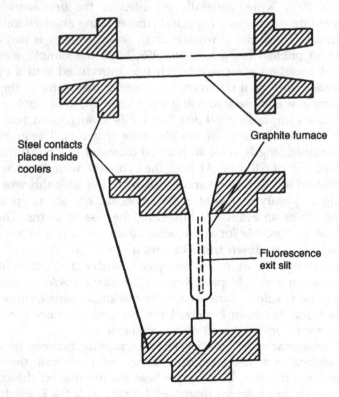

**Figure 3.2** Graphite furnace used for atomic fluorescence spectrometry.

cup for AFS are shown in Fig. 3.2. Using a power supply of 400 A at 10 V, the furnace could be heated to 2600 °C in a few seconds. Typical solution volumes of 5–200 mm³ were used.

Another type of atomizer was also developed and was popular for a while, but has since been abandoned. This was the **West rod** atomizer which was first reported in 1969. No tube was used; the sample was applied directly to an **electrically heated filament**. The graphite filament (2 mm diameter, 40 mm long), supported by water-cooled electrodes, could be heated to 2000–3000 °C within 5 s by the passage of a current of about 70 A at up to 12 V. Small liquid samples (1–5 mm³) were pipetted on to a depression on the rod. While the original filament was enclosed in an **inert gas** purged chamber, it was later found simpler to **shield** the graphite from the air by a simple flow of shielding gas around the filament. The apparatus is shown in Fig. 3.3. This apparatus was, however, prone to severe interference effects, although the greatest sensitivity and freedom from matrix effects was found when making observations of the atom cloud immediately above the filament. By **drilling holes** in the filament, a compromise

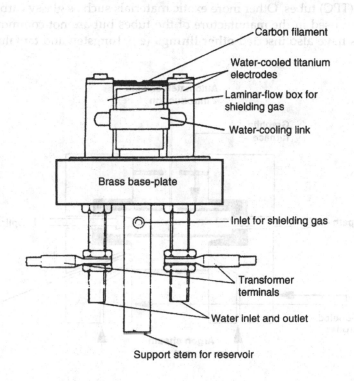

**Figure 3.3** West rod atomizer.

between the extended thermal contact of the furnace and the simplicity of the filament (or rod) has been reported.

## 3.2  HEATED GRAPHITE ATOMIZERS

Tubes are typically 20–30 mm long and 5–10 mm in diameter after the design of Massman. In the past, the tube may have been turned down at the centre to increase the temperature at that point, or the whole tube may have tapered towards the centre ('profiling') to shape the tube to the optical beam and increase the free atom density at the centre. Modern tubes tend not to have either of these modifications made to them. The graphite tube is held in place between two electrodes, axially in line with the light source, as shown in Fig. 3.4.

The two major disadvantages of graphite are its porosity and tendency for carbide formation. These may be partially overcome by coating the tube with **pyrolytic graphite** (e.g. by heating the tubes in a methane atmosphere), which is far less porous. Some manufacturers also produce **total pyrolytic carbon** (TPC) tubes. Other more exotic materials such as glassy carbon have also been used in the manufacture of the tubes but are not common. Some workers have also inserted **other linings** (e.g. tungsten and tantalum) into

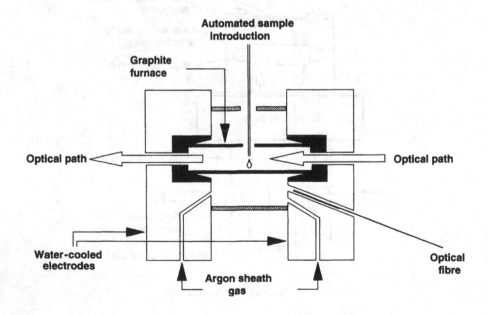

**Figure 3.4**  Modern design for a graphite furnace.

the furnace, or deposited a carbide lining on the inner wall (e.g. using lanthanum salts).

The furnace is heated by **low voltage** (usually 10 V) and **high current** (up to 500 A) from a well stabilized step-down transformer. For optimum precision, the voltage should be well stabilized, often by a feedback loop which may be temperature feed-back based (see Section 3.6.1). A rapid rise-time of the temperature is also preferable, because of theoretical considerations of peak shapes. This has implications for power supply design and furnace design, as will be shown below. Currently, furnaces are available that reach temperatures of up to 3000 °C, and temperatures of 2500 °C should be reached in less than 2 s in a well designed furnace.

The furnace is **purged** with an inert gas, usually **nitrogen or argon**. Argon, with a small addition of methane, is also used to provide continuous pyrolytic coating. There are some **chemical effects** between nitrogen and certain elements, e.g. titanium, vanadium and barium, (extremely refractory nitrides are formed) and the **rate of diffusion** of argon is less. This latter effect means that slightly larger signals are usually observed in argon. A gas flow that sweeps into the tube and out of the centre hole has been shown to reduce problems of background scatter. The flow may often be stopped during atomization to prevent dilution.

The whole atomizer may be **water cooled** to improve precision and increase the speed of analysis. The tube is positioned in place of the burner in an atomic absorption spectrometer, so that the light passes through it. Liquid samples (5–100 mm$^3$) are placed in the furnace, via the injection hole in the centre, often using an **autosampler** but occasionally using a **micro-pipette** with a disposable, 'dart-like' tip. **Solid samples** may also be introduced; in some designs, this may be achieved using special graphite boats. The sample introduction step is usually the main **source of imprecision** and may also be a source of **contamination**. The precision is improved if an autosampler is used. These samplers have been of two types: automatic injectors and a type in which the sample was nebulized into the furnace prior to atomization. This latter type was far less common.

The power supply controlling the furnace can be **programmed** so as to **dry** the sample after injection, **ash** it at an intermediate temperature (say 500 °C) and **atomize** it (Fig. 3.5). The temperature and duration of each of these steps can usually be controlled over a wide range. Optimizing the operating conditions of the furnace (or **'programming the furnace'**) is a vital step in the development of analytical methods. In the drying phase, the solvent must be driven off without problems of **'spitting'**. Drying of organic solvents tends to give particular problems, and the ashing conditions are most critical. It is essential to remove organic matter by **pyrolysis** and as many volatile components of the matrix as possible, but to avoid any loss of the analyte, either as the element or as a volatile salt, such as a halide.

**1. Add 50 µl of sample**

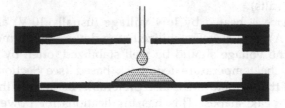

**2. Flush furnace with gas and dry droplet slowly**

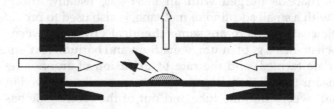

**3. Flush furnace with gas and ash the sample**

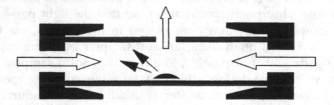

**4. Divert gas and vaporize analyte at high temperature**

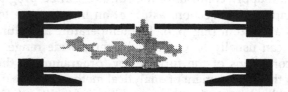

**Figure 3.5** Typical analytical heating cycle for a graphite furnace.

The atomization temperature is usually chosen so as to give a **rapid peak**. It should not be so hot as to damage the tube unnecessarily or distil off involatile contaminants, or so cool as to lose sensitivity or create memory effects (although a tube clean, i.e. a high-temperature cycle, can be included in the programme).

A spectrometer with **rapid response electronics** should be used for electrothermal atomization, as it must follow the **transient absorption event** in the tube. **Automatic simultaneous background correction** (see Section 2.2.5.2) is virtually essential, as non-specific absorption problems are very severe. It is important that the continuum light follows exactly the **same path** through the furnace as the radiation from the line source (assuming a deuterium lamp is being used rather than Smith–Hieftje or Zeeman effect). The time interval between the two source pulses should be as short as possible (a chopping frequency of at least 50 Hz) because of the transient nature of the signal.

Usually **peak area** of the transient absorption signal is measured as it provides greater precision. **Peak height** measurements may also be used, but may yield erroneous results if the analyte is in a different chemical form in the standards compared with the samples.

---

**Q.** What are the advantages of pyrolytic graphite coatings for furnaces?

---

**Q.** Why is argon preferred to nitrogen as the purge gas?

---

**Q.** Why are fast response electronics and background correction essential when using furnace atomization?

---

## 3.3 OTHER ATOMIZERS

Several other types of atomizer have been developed. Some of these are based on the design of the West rod, but others have made tubular atomizers from extremely refractory metals such as tungsten, tantalum and molybdenum. This latter class of atomizers tend to be made in-house by some laboratories and, at present, do not have any commercial suppliers. They have the advantage of being inert and non-porous so there is little interaction with the analyte, so that they can be used for the determination of elements which form refractory carbides. However, after extended use and in the presence of some acids, many of these atomizers become brittle and distorted.

Commercially available **tantalum filaments** used to be produced, but these too suffered similar disadvantages and have since ceased to be marketed.

The West rod atomizers were marketed for a while, but their production has also ceased. These atomizers had the advantages of **simplicity, low power** requirements (less than half that required by a furnace) and **fast heating rate** (2000 K s$^{-1}$). They were, however, considered to be extremely prone to interferences. This was attributed to the **rapid cooling** of the atoms once they had left the filament. This was partially overcome by setting the light beam so that it **grazed the surface** of the rod.

Other modifications that have been produced include incorporating a cavity or 'minifurnace' into the rod. This was accomplished by placing a **hollow-cathode cylinder** or a **graphite cup** between two spring-loaded rods. These mini-furnaces could cope with samples of up to 20 mm$^3$, or in some cases **solid** samples could also be analysed.

**Q.** What are the advantages and disadvantages of metal furnaces compared with graphite furnaces?

**Q.** What are the advantages and disadvantages of filaments compared with graphite furnaces?

**Q.** How has the original design of the West rod been modified to minimize these disadvantages?

## 3.4  ATOMIZATION MECHANISMS

An investigation of both thermodynamic and kinetic considerations is necessary in the understanding of atomization in graphite atomizers.

### 3.4.1  Thermodynamic considerations

Several possible reactions may be involved.

(i)  Conversion of **metal salts to the oxide.** When heated strongly after deposition from aqueous solution, nitrates, sulphates and some chlorides are usually converted to the oxide.

(ii)  **Evaporation of the metal oxide or metal halide** prior to atomization. Most **metal halides** are **volatile** and evaporate before atomization. Some metal oxides have **measurable vapour pressures** at the temperatures at which atomization is first observed to occur (the so-called '**appearance temperature**'). Using the simple **gas law**, 1 ng of a metal oxide (molecular weight $= 100$), completely vaporized into a volume of 100 mm$^3$ at 1000 K, would exert a vapour pressure of $6 \times 10^{-3}$ mmHg. Thus, unless this exceeds the saturated vapour pressure at 1000 K, complete evaporation could be expected.

(iii) **Thermal dissociation of the salt** or oxide. This will be related to the **temperature, pressure** and **other species** present, For dissociation of an oxide, the relevant equations are

$$-\Delta G = RT \ln K_p$$

and

$$\alpha = \frac{K_p}{K_p + P_{O_2}}$$

where $G$ is the free energy of the reaction, $\alpha$ is the degree of dissociation of the oxide, $K_p$ is the equilibrium constant for the dissociation and $P_{O_2}$ is the partial pressure of oxygen. The partial pressure of oxygen is effectively controlled by the equilibria

$$2C + O_2 \rightleftharpoons 2CO$$
$$2CO + O_2 \rightleftharpoons 2CO_2$$

(iv) **Reduction of the metal oxide.** Data concerning **carbon reduction** of metal oxides are readily available in forms such as **Ellingham diagrams** used by chemists and metallurgists. The essential reactions to consider are

$$MO(s/l) + C(s) \longrightarrow M(g) + CO(g)$$
$$C(s) + O_2(g) \longrightarrow CO_2(g)$$
$$2C(s) + O_2(g) \longrightarrow 2CO(g)$$

Such a thermodynamic approach can be extended to consider problems such as **carbide formation**:

$$MO + 2C \longrightarrow MC + CO$$

The weakness of this approach is that it deals with **equilibrium criteria,** whereas the situation in a furnace and certainly on a filament is highly dynamic. It must also assume some dissociation at all temperatures, and thus the appearance temperature becomes that at which the free metal is first detectable; hence the parameter should be dependent upon the detection limit and concentration. Useful insights have been afforded by the application of thermodynamics, but clearly kinetic factors must also play a role.

## 3.4.2 Kinetic considerations

As L'Vov first pointed out, to achieve analytically useful sensitivity, the **rate of formation** of the free atoms must be equal to or greater than their rate of

**removal** from the atom cell. If $N$ is the number of atoms at time $t$, $dN/dt$ is the rate of change of the number of atoms.

In a graphite atomizer, the atoms will appear according to a kinetic **rate equation** which will probably contain an exponential function. As the number of atoms in the atom cell increases, so does the rate of removal, until, at the **absorption maximum** (peak height measurement), the rate of formation equals the rate of removal. Thereafter, removal **dominates**.

The response function, which is normally a **peak** and may be distorted to some extent by the electronics, clearly is the **difference** between the formation and removal functions at that time. Atoms leave the atom cell partly by **diffusion** and according to the velocity of the **purge gas**. The rate of formation of atoms is more difficult to identify.

L'Vov first developed a kinetic model for atomization, based on **increasing temperature**, such as that found in a rod-type system. Fuller developed a model for atomization under **isothermal conditions**, applicable to less volatile elements in tube-type systems. Fuller's model assumes **first-order kinetics** and involves a number of other assumptions, but its usefulness has been demonstrated. For example, it confirms the usefulness of **integration** when atomization is slow (at relatively low temperatures or when investigating involatile elements), the enhancement of sensitivity available from **stopping the flow** of purge gas during the atomization cycle, and indicates methods for the control of interferences.

It is clear from experimental data that the rate of removal of the analyte can **exceed** the rate of supply. Hence there is an advantage to be obtained in rapid heating (e.g. $1000 \text{ K s}^{-1}$) and stopping the purge-gas flow during atomization.

While a full treatment of kinetic theories is beyond the scope of this book, it is clear that they have added much to our understanding of observed events in tube furnaces. It can be expected that more sophisticated models will offer more comprehensive explanations of observed behaviour.

---

**Q.** Would you expect the electrothermal atomization mechanism for zinc to differ depending on whether the sample was dissolved in nitric or hydrochloric acid?

---

**Q.** How can we explain the observed shapes of peaks obtained using furnace atomizers?

---

## 3.5 INTERFERENCES

Electrothermal atomizers are usually regarded as being **more prone** than flames to interferences, although it has now been clearly demonstrated that the nature and extent of interferences may vary in different types of atomizers. We shall now consider the problems encountered in the furnace type atomizers used in AAS.

## 3.5.1 Physical interferences

The steep thermal gradient along the tube means that any **variation in the sample position** (e.g. because of pipetting, or spreading due to surface tension and viscosity effects) will alter the atomization peak shape. **Peak area** integration will help to minimize this problem, as will a rapid heating ramp and isothermal operation (see Sections 3.6.2 and 3.6.3).

## 3.5.2 Background absorption

As already noted, the effects are usually severe. Large amounts of matrix are volatilized in a confined space. **Molecules** may exhibit **absorption spectra** in the region of interest; this is especially true of **alkali metal halides**. Particulate **smoke** also contributes to this problem. **Light emission** from the incandescent walls may further distort the baseline. Every effort should be made to reduce background absorption effects in method development, e.g. by attempting to reduce the matrix during the ash stage. **Background correction** should be applied routinely (see Section 2.2.5.2), remembering that background absorption is often at the 90% level.

## 3.5.3 Memory effects

**Incomplete atomization** of involatile elements can sometimes cause a problem. Often this can be overcome by firing a so-called 'cleaning' cycle at maximum power between analytical cycles.

## 3.5.4 Chemical interferences

### 3.5.4.1 Losses of analyte as a volatile salt

This is particularly likely to occur when halides are present, at the ashing or atomizing stage at temperatures too low to afford atomization, and can lead to losses of, for example, $CaCl_2$ or $PbCl_2$. The use of hydrochloric acid for sample dissolution should be avoided. If chloride is present, excess nitric acid can be added to the sample and hydrogen chloride **boiled off** during atomization. Preferably, 50% ammonium nitrate can be added (with care) to give the reaction

$$NaCl + NH_4NO_3 \longrightarrow NaNO_3 + NH_4Cl$$

| boils at | decomposes | decomposes | sublimes |
|---|---|---|---|
| 1413 °C | at 210 °C | at 380 °C | at 335 °C |

The compounds of Group V elements are often volatile, and loss of, for example, **arsenic, selenium and tellurium** during ashing of the sample can be reduced by the addition of **nickel**, to form nickel arsenide. Such stabilization procedures are called **matrix modification** (see Section 3.6.4).

### 3.5.4.2 Anion and cation interferences

Many of these have been reported, e.g. 0.1% v/v mineral acids interfere with several elements. While some insights are being gained in terms of the theories discussed above, interferences are usually combatted by matching standards and samples or by the method of standard additions.

### 3.5.4.3 Carbide formation

The apparent slow atomization of some elements may be caused by carbide formation. Rapid heating and a reproducible surface (e.g. a pyrolytic surface) help reduce this problem, as does coating of the tube (e.g. with lanthanum, using lanthanum nitrate solution), and the use of metallic tubes or boats.

### 3.5.4.4 Condensation

Some interferences appear to be **vapour-phase effects** and are presumably due to occlusion of analyte elements into particles of matrix. The use of **platforms** in furnaces (see Section 3.6.3), the use of **reactive purge gases** (e.g. hydrogen) and **dispersion** of the matrix, e.g. by using an organic acid such as ascorbic acid, can in some cases reduce such interferences.

---

**Q.** How can physical interferences be minimized?

---

**Q.** Is background correction more essential with flame or with electrothermal atomizers?

---

**Q.** How can interference from chloride ions be minimized?

## 3.6 METHODS OF OVERCOMING INTERFERENCES

### 3.6.1 Control of furnace temperature

The **faster** the rate of heating, the higher is the **density** of the atoms which will be formed in the **transient** atomic cloud. This leads to improved analytical **sensitivity**, provided that the electronics can follow the **rapid signals**. To obtain good reproducibility, it is necessary to **control the temperature** actually reached by the tube. While it is relatively simple to stabilize the **applied voltage**, variations in the **resistance** of individual tubes and the degradation of the tube as it is used mean that the temperature achieved must be **monitored**. Thus the voltage applied should be controlled via a **feed-back circuit**, linked to some method of sensing the tube temperature. This temperature may be measured using a **thermocouple**, which unfortunately may suffer from temperature lag, or more commonly by using a **light sensor**, e.g. an infrared sensor which views the tube-wall radiation via a fibre optic. In the latter case, it is clearly important that the end of the fibre optic remains clean. In a well designed modern furnace, therefore, the control settings 'dry', 'ash', 'atomize' and 'clean' should refer to **reproducible temperatures** (probably inaccurately known but precise) rather than to different applied powers.

### 3.6.2 The effect of the orientation of tube heating

Initially it was thought that maximum sensitivity would be obtained from larger tubes, as these would retain the atomic vapour for a longer period. Long tubes, however, require **more power** to achieve a given temperature, and **smaller tubes** can be heated more **rapidly**. Therefore, in the interests of rapid heating and **simplicity** of power supply, small tubes, typically of 20–30 mm length, are preferred. The tube should not be so short that the atoms **escape** from the tube and cool too rapidly; modern tubes give residence times of about 0.5 s.

Most atomizers are heated **longitudinally** i.e. they are clamped at either end by electrodes and the electricity is passed through the length of the tube. This leads to a temperature gradient along the tube, with the central portion being several hundred degrees hotter than the ends. This can lead to **condensation** of the analyte or **recombination** with other species at the cooler ends of the tube. Some modern instruments heat the tube **transversely**, i.e. from the sides. With this method of heating, a temperature gradient along the tube does not exist. Therefore, the efficiency

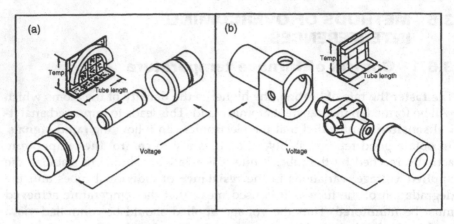

**Figure 3.6** Temperature distribution in (a) longitudinally and (b) transversely heated graphite furnace.

of atomization is increased, there is less tailing of the absorption signal, there is a substantial decrease in the carry-over (or memory) effect and refractory analytes may be determined more readily. The effects of the different modes of heating are shown diagrammatically in Fig. 3.6.

### 3.6.3 Isothermal operation

The early L'Vov design had the advantage that the sample was atomised into a **hot environment** (so-called isothermal operation). In the Massmann-style furnace, the atoms form as soon as the temperature of the tube wall reaches **atomization temperature**. The gas within the tube will be somewhat cooler so when the atoms leave the tube wall, they immediately cool and may then condense or recombine with other matrix components and will then be lost analytically. This leads to a variety of possible interferences which for many years caused serious problems. L'Vov then suggested a simple device that gives many of the advantages of isothermal atomization while essentially retaining the simplicity of the Massmann design. The sample is pipetted on to a small graphite **platform** only loosely connected to the tube walls (Fig. 3.7). This platform is then heated, partly by radiative and convective means, and atomization occurs only when the **surrounding gas** is relatively hot. This device is sometimes referred to as the **L'Vov platform** and miniature graphite plates can be purchased for this purpose. Alternatively, an old tube may be broken to provide fractions of the tube about one quarter of the circumference wide and 5 mm long. Some more

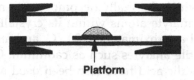

**Platform**

**Figure 3.7** Isothermal operation using a platform.

modern instruments even have a platform as an integral part of the tube. Using such platforms, some **interference effects** can be reduced noticeably.

More recently, an alternative technique has been developed. The use of **probe atomization** became popular in the mid-1980s and has been shown to offer the same advantages as a platform. The sample is pipetted on to a graphite probe and the normal drying and ashing cycles are performed. The probe is then removed from the tube, which is then heated to the atomization temperature. When this has been done, the probe is re-introduced into the tube and is heated by the hot gas present, allowing the atoms to form in an isothermal atmosphere.

## 3.6.4 Matrix modification

Matrix modification is a term first coined by Ediger in 1975 [*At. Absorpt. Newsl.* **14**, 127 (1975)], although 'chemical modification' is now often the preferred term. It is a process by which a chemical or a combination of chemicals is added to the sample so that the analyte may be separated from the matrix more easily, hence facilitating interference-free determinations. This may be achieved in two ways. The first is to accelerate the removal of the matrix by adding a chemical that volatilizes the matrix at a lower temperature. Examples of agents such as these include ammonium nitrate, nitric acid or oxygen (air). Ammonium nitrate and nitric acid assist in the removal of chloride ions (see Section 3.5.4) whilst leaving the analyte in the atomizer. Oxygen or air is used frequently if biological samples are to be analysed. Gases such as these literally combust the organic matrix and leave the analyte in the atomizer ready to be atomized into a relatively matrix-free (and hence interference-free) atmosphere.

A large number of matrix modifiers have been developed that **thermally stabilize** the analyte, allowing higher ash temperatures to be used without analyte loss. In this way, more matrix may be removed leaving less to interfere with the analyte's determination. Examples of this type of matrix modifier include some transition metal ions, e.g. Ni and Pd, which form thermally stable intermetallic compounds with the metalloids, e.g. As–Ni, magnesium nitrate, which thermally decomposes to magnesium oxide and in the process **traps** analyte atoms within its crystalline matrix, thermally stabilizing them until approximately 1100 °C, and ammonium phosphate, which stabilizes volatile analytes such as cadmium. Organic acids such as ascorbic, citric and oxalic acid have also been used as reducing agents for analytes such as lead. Often, a combination of modifiers is used. One of the most common combinations is that of magnesium nitrate and palladium nitrate. This has been termed the **universal modifier** because of its use for a large number of analytes.

### 3.6.5　The STPF concept

The **stabilized temperature platform furnace (STPF)** concept was first devised by Slavin *et al.* It is a collection of recommendations to be followed to enable determinations to be as free from interferences as possible. These recommendations include (i) isothermal operation; (ii) the use of a matrix modifier; (iii) an integrated absorbance signal rather than peak height measurements; (iv) a rapid heating rate during atomization; (v) fast electronic circuits to follow the transient signal; and (vi) the use of a powerful background correction system such as the Zeeman effect. Most or all of these recommendations are incorporated into virtually all analytical protocols nowadays and this, in conjunction with the transversely heated tubes, has decreased the interference effects observed considerably.

**Q.** What are the advantages of transversely heating the graphite tube?

**Q.** Why is isothermal operation recommended for overcoming interferences?

**Q.** Explain why matrix modification may lead to more efficient matrix removal.

## 3.7　OTHER ELECTROTHERMAL TECHNIQUES

The interest in the graphite furnace as an emission source stems from the desire to achieve a multi-element detection capability whilst retaining the low limits of detection and the possibility of sample pretreatment. Several approaches have been made, many of which have now been superseded. The detection limits of the technique were, however, comparable to those obtainable by electrothermal AAS (ETAAS). Excitation sources that rely on non-thermal processes are often preferable to those that rely on thermal processes because of their low background emission. Two such techniques are discussed briefly below.

### 3.7.1 Furnace atomic non-thermal excitation spectrometry (FANES)

The procedure used for this technique is similar to that for ETAAS. The sample is pipetted into the graphite tube and dried and ashed in the normal manner. The system is then evacuated, before allowing argon to partially re-pressurize it. A glow discharge is then generated and the sample atomized into it by heating the graphite tube. Once in the discharge, the atoms become excited and give off light, the intensity of which is measured. The technique, although popular for a while, is complex and hence inconvenient, because of the necessity to work at low pressure (< 200 Torr). A few research papers are still published each year but, as yet, the technique has not found universal acceptance.

### 3.7.2 Furnace atomization plasma emission spectrometry (FAPES)

By coupling an RF generator to a graphite furnace containing a central tungsten electrode, an atmospheric pressure plasma within the tube is formed. Helium is usually used rather than argon as the sheath and internal purge gas. The sample is pipetted into the tube, dried and ashed in the normal way and then an RF power of typically 50 W is applied. The plasma formed around the central tungsten electrode is allowed to burn for 20 s, and then the high temperature atomization cycle of the programme is initiated. The atoms pass into the plasma and the light they emit is measured. The RF power is then turned off and, after the normal cooling stage, the entire sequence may be repeated.

---

**Q.** What are the advantages and disadvantages of these alternative techniques when compared with conventional electrothermal atomization?

---

## 3.8 APPLICATIONS

Electrothermal atomization is particularly useful when the amount of sample is very **small**, when very **low levels** of detection are required and when the matrix is **dilute** or **volatile**. These criteria often apply to **clinical samples** (a pin-prick sample of blood produces only 50–100 mm$^3$ of whole blood, but this is sufficient for analysis using an electrothermal atomizer, hence it is not essential for an intravenous sample to be taken). For such samples, often pretreatment is not required, and body fluids and biological tissues can be **ashed** *in situ* in the furnace. This also applies to some **foods**, although others may need some preliminary **wet ashing**.

**Oils** can be injected **directly** or in a **dilute** form, e.g. diluted with xylene. Organometallic standards are recommended.

**Metallurgical** samples are perhaps not as amenable to electrothermal atomization as some types of sample, but as high sensitivity is often only required for the most volatile elements, useful information can be obtained. Problems may be encountered from chloride when using aqua regia to dissolve samples. An interesting application is the placing of weighed **solid samples** directly into the furnace for ultra-trace analysis of volatile elements. This procedure is, however, notoriously imprecise, as the accurate weighing of samples of less than 10 mg is difficult.

**Waters** are the subject of a voluminous literature, and various methods have been proposed to overcome some of the interferences encountered, e.g. by adding ascorbic acid or lanthanum to remove interferences when determining lead in hard water. **Saline waters** present particular problems (e.g. from background absorption), and a preliminary separation may be advisable.

**Air particulates** are usually dissolved before analysis, but again, solid samples (e.g. on glass-fibre filters) have been analysed directly in furnaces.

The **standard addition** method of **calibration** (see Chapter 1) is often used to combat the uncertainties of varying interference effects in electrothermal atomization. However, care should be taken with this approach, as errors from spurious **blanks** and **background** may go undetected. It must also be emphasized that the technique of standard additions does not correct for all types of interference.

The **literature** on applications of electrothermal atomizers is now extremely large and, because of the details of the furnace programmes used, is well worth consulting. The tables in the Atomic Spectrometry Updates reports (see Appendix C) offer the best way of accessing this information.

---

**Q.** Why is electrothermal atomization widely used in clinical applications?

---

**Q.** What are the advantages and disadvantages of the standard addition method for the above applications?

---

## 3.9 THE RELATIVE MERITS OF ELECTROTHERMAL ATOMIZATION

It is probably best to regard electrothermal and flame atomization and plasma-based instruments as complementary techniques. Some factors which govern the choice of technique for a given application are given below.

### 3.9.1 Advantages of electrothermal atomization

(i) **Increased sensitivity**: the **theoretical improvement** obtainable in **electrothermal atomization** in comparison with flame atomization has been calculated by several workers. Such calculations are based on the **poor nebulization efficiency** associated with flames (10%), the **rapid dilution in the flame** with the expansion of the flame gases and the **short residence time**. Improvements in detection limits of furnaces compared with flames range up to 4000-fold for zinc and are typically in the range 100–1000-fold. For ICP-AES instruments, the nebulization efficiency is even worse (2%) and the other shortcomings listed above are equally applicable. However, the technique cannot compete with ICP-MS in terms of sensitivity.

(ii) **Decreased sample size**: the minimum requirement of a flame or a plasma instrument is 500 mm³, except where flow injection or pulse nebulization is used. For electrothermal atomization, sample sizes of 1–100 mm³, typically 20–30 mm³, are used. This means that dramatic **improvements in absolute sensitivities** are obtained and measurements of picogram amounts of analyte are possible. Thus, electrothermal atomization offers absolute analytical sensitivity comparable to that associated with neutron activation analysis. Electrothermal atomization has particular advantages in situations where sample size is limited. Table 3.1 shows typical characteristic masses obtainable in electrothermal atomization. The use of limit of detection is rarely used for this technique because it depends so heavily on the experimental conditions and on the injection volume.

(iii) *In situ* **sample treatment**: often tedious ashing procedures can be avoided by judicious choice of acids and ashing temperature in the furnace.

(iv) Direct analysis of **solid samples**: solid samples can be placed directly in or on electrothermal atomizers, often using purpose-made accessories. The only ways to analyse solids in plasma instruments is

**Table 3.1**　The characteristic masses of some analytes in ETA-AAS.

| Element | Characteristic mass (pg) | Element | Characteristic mass (pg) |
|---------|--------------------------|---------|--------------------------|
| Ag | 0.9 | Fe | 1.5 |
| Al | 5.0 | Mn | 0.7 |
| As | 6.0 | Mo | 8.0 |
| Au | 4.0 | Ni | 5.0 |
| Ba | 12.0 | Pb | 2.5 |
| Cd | 0.25 | Se | 20.0 |
| Co | 4.0 | Ti | 45.0 |
| Cr | 1.5 | V | 20.0 |
| Cu | 2.5 | | |

      to use laser ablation, or to introduce the sample by electrothermal vaporization or direct insertion.

(v) **Cheapness of operation**: there is a low consumption of argon, graphite tubes and electricity. This compares favourably with the consumption of gases by a flame or a plasma instrument.

(vi) **Safety** of operation: explosive gases and flames may be avoided, less toxic fumes are produced, flame products are absent and smaller samples used. Enclosed use means that radioactive samples may be handled.

(vii) **Suitability** for working in the **vacuum ultraviolet region of the spectrum**: argon does not absorb in the vacuum ultraviolet region, whereas flame gases do. Plasma instruments do not suffer from this problem provided that a suitable monochromator is used.

(viii) **Unattended operation**: the use of an autosampler means that unattended, overnight operation is possible, although most plasma instruments also have this facility.

## 3.9.2　Disadvantages of electrothermal atomization

(i) **Time**: a typical programme cycle for electrothermal atomization may take 2 min, whereas a flame or plasma determination typically takes 10 s. Work has been performed that has attempted to decrease the analysis time, e.g. by hot injection (where the sample is injected into a hot furnace to decrease the drying time) and the use of extremely short ashing times. In addition, the lack of a continuous reading makes setting-up more time consuming.

(ii) **Poor precision**: most imprecision in electrothermal atomization is associated with sample introduction, especially when manual pipetting is used. It cannot be expected that discrete signals can offer precision similar to that of integrated continuous signals. It should be noted that the precision for electrothermal atomization is typically 5% compared with 2–3% from a flame or plasma instrument.

(iii) **Interferences**: Electrothermal atomizers still suffer from more interferences than the nitrous oxide–acetylene flame. These interferences, however, have been reduced substantially over the last 10 years.

(iv) **Expense**: a good electrothermal atomizer with autosampler is an expensive instrument, costing £50 000–60 000, whereas a simple, very basic, flame spectrometer may cost as little as £10 000 (more normally £15 000–20 000). An ICP-AES instrument will cost in excess of £60 000 and ICP-MS instrument between £150 000 and £250 000.

(v) **Complicated programmes**: optimizing the conditions of electrothermal atomization is far more complicated than for a flame. ICP-AES and ICP-MS are more complicated instruments to operate *per se*, although it is easier to find compromise operating conditions.

(vi) **Small samples**: the small samples necessary in electrothermal atomization present problems in sample handling and with homogeneity.

(vii) **Single element**: traditionally it has only been possible to determine one element at a time, whereas ICP-AES and ICP-MS are simultaneous multi-element techniques. However, new multi-element spectrometers are now available.

While it is to be expected that the effects of these disadvantages will continue to diminish as more becomes known about electrothermal atomization, currently it can be said that if there is sufficient sample for flame or ICP analysis, and that these techniques offer sufficient sensitivity, then they should be used in preference. Plasma techniques should be used in preference to the flame if more than one analyte is to be determined. Recently a multi-element, simultaneous electrothermal instrument has been developed. These spectrometers still use a suite of hollow cathode lamps as sources. At present, a maximum of six analytes can be determined simultaneously. This area is likely to expand very rapidly, which may lead to a resurgence in the technique. If the sensitivity of a flame or ICP-AES is insufficient, and ICP-MS cannot be afforded, electrothermal atomization comes into its own, and is invaluable when either high sensitivity is required or when only small amounts of sample are available.

**Q.** What are the particular advantages of electrothermal atomization?

**Q.** Would you use electrothermal atomization or some other technique for the following determinations: (1) zinc in a trade effluent at the 0.1 µg ml$^{-1}$ level; (2) cadmium in blood at the ng ml$^{-1}$ level; (3) lead in steel at (a) the 0.1% w/w level and (b) the 0.001% level?

# 4 PLASMA ATOMIC EMISSION SPECTROMETRY

## 4.1 THEORY

### 4.1.1 Atomic transitions

The **probability** of transitions from given energy levels of a fixed atomic population (e.g. between the lower level $i$ and upper level $j$) was expressed by **Einstein** in the form of three **coefficients**. These are termed **transition probabilities** as follows:

$$A_{ji} \quad \text{spontaneous emission} \quad j \longrightarrow i$$

$$B_{ij} \quad \text{absorption} \quad i \longrightarrow j$$

$$B_{ji} \quad \text{stimulated emission} \quad j \longrightarrow i$$

Such transitions are illustrated in Fig. 4.1, and they can be considered as representing the ratio of the number of atoms undergoing a transition to the number in the initial level. The intensity $I_{em}$ of a **spontaneous emission** line is related to $A_{ji}$ by the equation.

$$I_{em} = A_{ji} h \nu_{ji} N_j \tag{4.1}$$

When a system is in thermodynamic equilibrium the level population, i.e. the number of atoms $N_j$ in the excited state, is given by the **Boltzmann distribution law**:

$$N_j = N_0 \frac{g_j}{g_0} \exp[-(E_j - E_0)/kT] \tag{4.2}$$

where $N_0$ is the number of atoms in the ground (unexcited) state with an energy $E_0 = 0$, $g_j$ and $g_0$ are the statistical weights of the $j$th (excited)

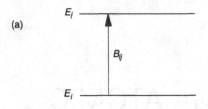

(a)

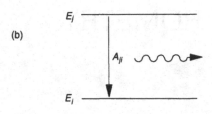

(b)

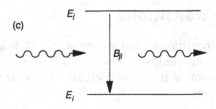

(c)

**Figure 4.1**  Schematic representation of atomic transitions between electronic states.

and ground states, respectively (where $g = 2J + 1$, $J$ is the third quantum number), $k$ is the Boltzmann constant and $T$ is the temperature. Thus,

$$\frac{N_j}{N_0} = \frac{g_j}{g_0} \frac{\exp[-(E_j/kT)]}{\exp[-(E_0/kT)]} \tag{4.3}$$

If we express $N$, the total number of atoms present, as the sum of the population of all levels, i.e. $N = \Sigma_j N_j$, then

$$\frac{N_j}{N} = \frac{g_j \exp[-(E_j/kT)]}{\sum_j g_j \exp[-(E_j/kT)]}$$

$$= \frac{g_j \exp[-(E_j/kT)]}{F(T)} \tag{4.4}$$

where $F(T)$ is known as the **partition function**.

If self-absorption is neglected for a system in thermodynamic equilibrium:

$$I_{em} = A_{ji} h \nu_{ji} \frac{N g_j \exp[-(E_j/kT)]}{F(T)} \quad (4.5)$$

A similar result is more readily, if less rigorously, obtained if we assume that virtually all the atoms remain in the ground state (the strength of this assumption can be seen in Table 4.1). Thus, Eqn. 4.2 becomes

$$N_j = N \frac{g_j}{g_0} \exp[-(E_j/kT)] \quad (4.6)$$

and Eqn. (4.1) becomes

$$I_{em} = A_{ji} h \nu_{ji} N \frac{g_j}{g_0} \exp[-(E_j/kT)] \quad (4.7)$$

This is similar to Eqn. 4.5 for practical purposes and the reader may prefer this simplified derivation.

Thus, the intensity of atomic emission is critically dependent on the **temperature**. It also follows that when low concentrations of analyte atoms are used (i.e. when self-absorption is negligible), the plot of emission intensity against sample concentration is a **straight line**.

## 4.1.2 Broadening

The result of a radiative atomic transition from an upper to a lower energy level is radiation at a particular wavelength, as defined by

$$\lambda = \frac{hc}{E_j - E_0} \quad (4.8)$$

where $h$ is Planck's constant and $c$ is the velocity of light *in vacuo*.

However, atomic lines are not infinitely thin as would be expected and their width is discussed by talking about half-width ($\Delta \nu$ cm$^{-1}$), illustrated in Fig. 4.2a.

**Natural broadening** occurs because of the finite lifetime ($\tau$) of the atom in the excited state. Heisenberg's uncertainty principle states that if we know the state of the atom, we must have uncertainty in the energy level. We assume that $\tau$ for the ground state is infinity and therefore for a resonance line the natural width $\Delta \nu_N = \frac{1}{2}\pi \tau$.

**Doppler broadening** arises from the random thermal motion of the atoms relative to the observer. The velocity $V_x$ of an atom in the line of sight will vary according to the Maxwell distribution, the atoms moving in all directions relative to the observer. The frequency will be displaced by

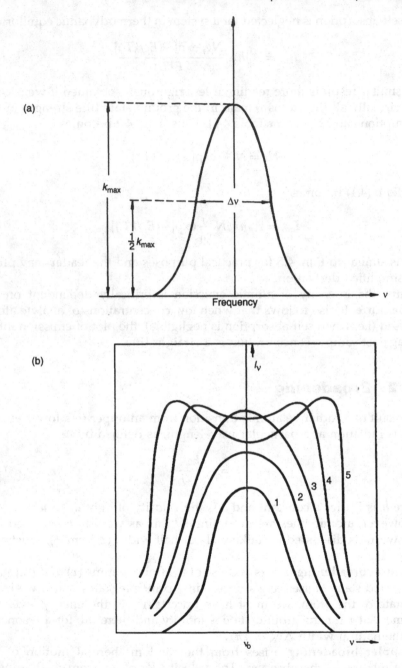

**Figure 4.2** Profile of an atomic line: (a) the half width $\Delta \nu$ is the width of the line when $k_\nu = 1/2 k_{max}$; (b) the effect of self-absorption as the concentration of atoms increases from 1 to 5.

$\Delta v = (V_x/c)v_0$. To this is applied the distribution of velocities. After evaluation of constants, this simplifies to the Doppler half-width:

$$\Delta v_D = 7.16 \times 10^{-6} v_0 \sqrt{(T/M)} \qquad (4.9)$$

where $M$ = relative atomic mass. The Doppler effect acts primarily on the centre of the profile.

**Lorentz (collisional) broadening** arises from collisions of atoms with atoms or molecules of a different kind. It has been shown experimentally that these collisions shift, broaden and cause asymmetry in the line. Different gases have different effects. Collisional theory offers the best fit equations to describe these events at the line centre, and statistical theory describes the events at the wings. Lorentz broadening increases with pressure ($P$) and temperature ($T$), and is generally regarded as being proportional to $P$ and $\sqrt{T}$. Thus, $\Delta v$ increases with increasing $T$ and $P$. It is accepted that Lorentz broadening affects the wings of the profile.

The profile of the line can be summarized by the Voigt profile:

$$K_v = K_0^{(D)} \frac{a}{\pi} \int_{-\infty}^{\infty} \frac{\exp(-y^2)\, dy}{a^2 + (\omega - y)^2} \qquad (4.10)$$

where

$$a = \frac{\Delta v_L}{\Delta v_D} \sqrt{\ln 2}, \; \omega = \frac{2(v - v_0)}{\Delta v_D} \sqrt{\ln 2} \text{ and } y = \frac{2\delta}{\Delta v_D} \sqrt{\ln 2}$$

$\Delta v_L$ is the Lorentz half-width and $\delta$ is the frequency displacement $v - v_0$.

**Stark broadening** occurs in the presence of an electric field, whereby the emission line is split into several less intense lines. At electron densities above $10^{13}$ the field is relatively inhomogeneous, the splitting is different for different atoms and the result is a single broadened line.

Other broadening processes also exist. **Holtsmark (resonance) broadening** arises from collisions between atoms of the same kind and is therefore negligible when compared with other collisions.

For resonance lines, self-absorption broadening may be very important, because it is applied to the sum of all the factors described above. As the maximum absorption occurs at the centre of the line, proportionally more intensity is lost on self-absorption here than at the wings. Thus, as the concentration of atoms in the atom cell increases, not only the intensity of the line but also its profile changes (Fig. 4.2b) High levels of self-absorption can actually result in self-reversal, i.e. a minimum at the centre of the line. This can be very significant for emission lines in flames but is far less pronounced in sources such as the inductively coupled plasma, which is a major advantage of this source.

**Q.** What determines the likelihood of spontaneous emission from an excited state?

**Q.** Which law describes the population of excited electronic states?

**Q.** Why are atomic emission lines not actually 'lines' in practice?

## 4.2 EXCITATION SOURCES

### 4.2.1 Flame sources

In the past, flames used for atomic absorption spectrometry have also been used for atomic emission spectrometry, and these are described in some detail in Chapter 2. However, the advent of plasma excitation sources has resulted in the demise of flame atomic emission spectrometry, for the reasons discussed in Section 4.2.3.

### 4.2.2 Plasma sources

A plasma is defined in the Oxford English Dictionary as

... a gas of positive ions and free electrons with an approximately equal positive and negative charge.

From the perspective of the atomic spectroscopist, desirable properties of plasmas include high thermal temperature and sufficient energy to excite and ionize atoms which are purposefully introduced for the purposes of analysis. In terms of atomic spectrometry, this means that we would generally wish to measure the absorption or emission of radiation in the near-ultraviolet (180–350 nm) and visible (350–770 nm) parts of the spectrum. In this sense, plasmas have been variously described as 'electrical flames' or 'partially ionized gases'. A working definition for atomic spectrometry could be as follows:

A plasma is a partially ionized gas with sufficiently high temperature to atomize, ionize and excite most of the elements in the Periodic Table.

A variety of gases such as argon, helium and air can be used to form analytically useful plasmas. However, argon and helium have been used most extensively because they can be obtained in a relatively pure form, have good characteristics for atomization, ionization and excitation of the analyte and are readily available, although expensive, throughout most of the world.

## 4.2.3   Flames vs plasmas

The effect of temperature and $E_j$, the energy difference between the excited and ground states, is best illustrated by the practical examples in Table 4.1. It can clearly be seen that the number of excited atoms, and hence the intensity of emission, increase very rapidly with increasing **temperature**, and that the number of excited atoms is greater the lower is the **energy level**.

The number of excited atoms at typical flame temperatures (ca 2200–3200 K) is very low indeed compared with the number of ground-state atoms, even for easily excited lines. For difficult-to-excite lines (e.g. Zn 213.9 nm), it can be shown that only about one excited atom will exist at any given time in an air–propane flame when aspirating a 1 mg l$^{-1}$ zinc solution. This is one reason why flames are poor sources for atomic emission spectrometry, but are well suited to atomic absorption spectrometry, i.e. most of the atoms are in the ground state. As will be seen, the typical temperatures obtainable in plasma sources are of the order of 8000 K, at which there is a much high ratio of excited-to ground-state atoms, and hence a much greater intensity of atomic emission.

### 4.2.3.1   Self-absorption

Figure 4.3 shows a **Grotrian diagram**, or partial energy level diagram, for sodium emission. If the oscillator strengths of the lines were equal (which, of course, they are not), we would expect to see maximum emission from lines where the upper excited states lie closest to the ground state, that is, where the excitation energy $E_j$ is small and the Boltzmann distribution predicts a greater population of excited atoms than at other states. The 3P–3S transition for sodium is very intense — the famous D lines — and hence it is an element very favourable for analysis by atomic emission spectrometry.

In a flame, as the concentration of atoms increases, the possibility increases that photons emitted by excited atoms in the hot region in the centre will collide with atoms in the cooler outer region of the flame, and thus be absorbed. This **self-absorption** effect contributes to the characteristic **curvature** of atomic emission calibration curves towards the concentration axis, as illustrated in Fig. 4.4. The inductively coupled plasma (ICP) tends to be hotter in the outer regions compared with the centre — a property known as **optical thinness** — so very little self-absorption occurs, even at high atom concentrations. For this reason, curvature of the calibration curve does not occur until very high atom concentrations are reached, which results in a much greater **linear dynamic range**.

**Table 4.1** Ratio of excited-to ground-state atoms for lines with varying excitation energies at increasing temperature.

| Element | Line (nm) | $g_i/g_0$ | $E_i$ (J) | Ratio of excited- to ground-state atoms | | | | | | |
|---|---|---|---|---|---|---|---|---|---|---|
| | | | | 2000 K | 3000 K | 4000 K | 5000 K | 6000 K | 7000 K | 8000 K |
| Cs | 852.1 | 2 | $2.34 \times 10^{-19}$ | $4.19 \times 10^{-4}$ | $7.06 \times 10^{-3}$ | $2.90 \times 10^{-2}$ | $6.75 \times 10^{-2}$ | $1.19 \times 10^{-1}$ | $1.78 \times 10^{-1}$ | $2.41 \times 10^{-1}$ |
| Na | 589.1 | 2 | $3.38 \times 10^{-19}$ | $9.65 \times 10^{-6}$ | $5.71 \times 10^{-4}$ | $4.39 \times 10^{-3}$ | $1.49 \times 10^{-2}$ | $3.38 \times 10^{-2}$ | $6.05 \times 10^{-2}$ | $9.37 \times 10^{-2}$ |
| Ca | 422.7 | 3 | $4.66 \times 10^{-19}$ | $1.40 \times 10^{-7}$ | $3.88 \times 10^{-5}$ | $6.47 \times 10^{-4}$ | $3.50 \times 10^{-3}$ | $1.08 \times 10^{-2}$ | $2.41 \times 10^{-2}$ | $4.41 \times 10^{-2}$ |
| Zn | 213.9 | 3 | $9.13 \times 10^{-19}$ | $1.30 \times 10^{-14}$ | $7.99 \times 10^{-10}$ | $1.98 \times 10^{-7}$ | $5.40 \times 10^{-6}$ | $4.90 \times 10^{-5}$ | $2.36 \times 10^{-4}$ | $7.70 \times 10^{-4}$ |

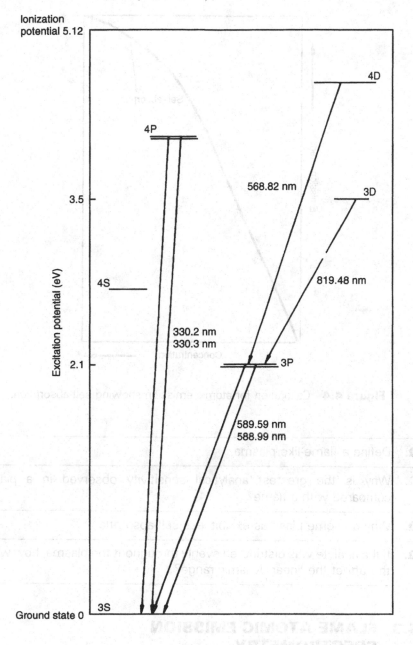

**Figure 4.3** Grotrian diagram for sodium emission.

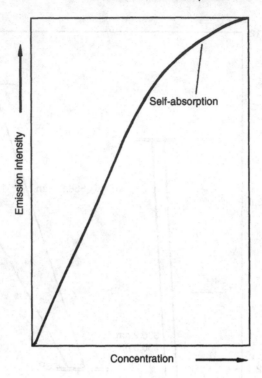

**Figure 4.4**  Calibration for atomic emission showing self-absorption.

---

**Q.**  Define a flame-like plasma.

---

**Q.**  Why is the greatest analytical sensitivity observed in a plasma compared with a flame?

---

**Q.**  Why do some plasmas exhibit less self-absorption?

---

**Q.**  If the analyte was distributed evenly throughout the plasma, how would this affect the linear dynamic range?

---

## 4.3  FLAME ATOMIC EMISSION SPECTROMETRY

A **wide range** of instrumentation may be employed for flame atomic emission spectrometry, from filter photometers to highly sophisticated instruments. The simplest instrument is a **flame photometer**, which is frequently used for alkali metal determinations in clinical and agricultural analysis. A low-temperature (e.g. air–natural gas) flame is used; thus only the most prominent lines are excited. These lines are isolated by coloured glass or interference **filters** (usually labelled K, Li, Na, etc.). Detection has typically been performed with a **barrier layer photocell**[†], **vacuum phototube, photomultiplier** or **solid-state detector**.

In the past, much atomic emission work has been performed on atomic absorption instruments which use a flame as the excitation source. However, these have been surpassed by instruments which utilise a **high-temperature plasma** as the excitation source, owing to their high sensitivity and increased linear dynamic range.

## 4.4 INDUCTIVELY COUPLED PLASMA ATOMIC EMISSION SPECTROMETRY

### 4.4.1 Plasma generation

Many different plasma sources exist, but by far the most common is the **inductively coupled plasma** (ICP). The ICP is generated by coupling the energy from a **radiofrequency** generator into a suitable gas via a **magnetic field** which is induced through a two- or three-turn, water-cooled copper coil. The radiofrequency energy is normally supplied at a frequency of **27.12 MHz**, delivering forward power at between 500 and 2000 W. Two gas flows, usually argon, flow in a **tangential** manner through the outer tubes of a concentric, three-tube quartz torch which is placed axially in the copper coil (Fig. 4.5). Because the **outer** and **intermediate gases** flow **tangentially** (i.e. they swirl around as they pass through the torch), the plasma is continually revolving and has a '**weak spot**' at the centre of its base, through which the **inner gas** flow, containing the sample, can be introduced. When the gas is seeded with electrons, usually by means of a spark, the electrons accelerate in the magnetic field and reach energies sufficient to ionize gaseous atoms in the field. Subsequent **collisions** with other gaseous atoms causes further ionization and so on, so that the plasma becomes **self-sustaining**. This occurs almost instantaneously. The magnetic field causes the ions and electrons to flow in the horizontal plane of the coil, thereby heating the neutral argon by **collisional energy exchange**, and a hot fireball is produced. The hottest part of the ICP has a temperature between **8000**

---

[†] A plate on which a thin layer of a semiconductor (e.g. selenium) has been deposited. A very thin transparent layer of silver is sputtered over the selenium to act as a collector electrode. Light falling on the semiconductor surface excites electrons which are released to the collector electrode. The current thus generated is measured using a galvanometer.

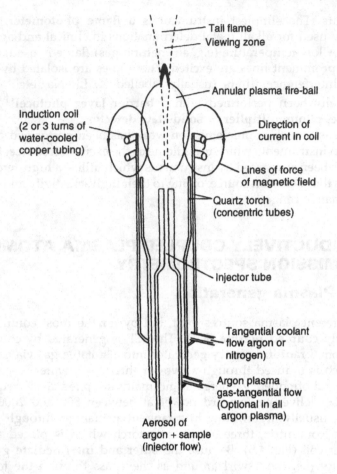

Tail flame

Viewing zone

Annular plasma fire-ball

Induction coil
(2 or 3 turns of
water-cooled
copper tubing)

Direction of
current in coil

Lines of force
of magnetic field

Quartz torch
(concentric tubes)

Injector tube

Tangential coolant
flow argon or
nitrogen)

Argon plasma
gas-tangential flow
(Optional in all
argon plasma)

Aerosol of
argon + sample
(injector flow)

**Figure 4.5**  Schematic view of an ICP.

and 10000 K, which is the temperature of the surface of the Sun, though the analytically useful region is in the **tail-flame** with a temperature between **5000 and 6000 K**.

If we confine ourselves to a discussion of an **argon ICP** then we can say that it consists mainly of the following species:

| | |
|---|---|
| Ar | neutral argon |
| $Ar^+$ | ionized argon |
| $Ar^*$ | excited argon |
| $Ar^{+*}$ | excited, ionized argon |
| $Ar^m$ | metastable argon |
| $e^-$ | electron |

In the absence of analyte atoms, water and sample matrix components, the predominant species will be Ar, $Ar^+$ and $e^-$, although the others are important when considering analyte ionization mechanisms. The RF energy used to sustain the plasma is only coupled into the outer region of the plasma, so these species are primarily formed in this region and are thermally transferred to the centre. This is known as the **skin effect**. The depth to which the RF energy will couple into the plasma gas is called the **skin depth**, and is determined by the frequency of the RF energy and the nature of the plasma gas. One desirable consequence of the skin effect is that the ICP is much more energetic in the outer region, which makes it **optically thin**. This means that emission from the centre will not be re-absorbed by unexcited atoms in the outer regions, such as occurs with flames. This lack of self-absorption means that ICP-AES has a large linear dynamic range.

The various species present in the plasma will circulate between the skin and the central region. Additionally, when the sample is introduced through the **axial channel**, the centre will contain a flow of cooler gas containing the analyte and species derived from the sample matrix and water. The region between the central channel and the plasma is known as the **boundary region**. The ICP can be divided axially into a number of other regions shown in Fig. 4.6. The region within the copper load coil is termed the **initial radiation zone** (IRZ) and can be identified by aspirating a solution of 1000 µg $ml^{-1}$ yttrium into the plasma, the intense **red atomic Y emission** indicating the IRZ. The third region is located between **10 and 20 mm** above the load coil and is called the **normal analytical zone** (NAZ). This is the zone which is normally observed for routine analytical determinations.

It is obvious from the foregoing the ICP exhibits a large degree of **spatial inhomogeneity**. In addition, the ICP is not in **thermal equilibrium** (TE) because the various collisional processes which occur in the plasma such as ionization, recombination, excitation and de-excitation are not in equilibrium. However, the ICP is thought to *approach* thermal equilibrium, a condition termed **local thermal equilibrium** (LTE). One consequence of this is that the ICP cannot be characterized by a single equilibrium temperature, with the **ionization temperature** ($T_{ion}$), **gas temperature** ($T_{gas}$), **excitation temperature** ($T_{exc}$) and **rotational temperature** ($T_{rot}$) all having different values. It may be confusing when one is used to regarding temperature as having a single value, until one remembers that it is impossible to measure the temperature of the ICP using everyday means (such as a thermometer) because it is simply too hot. Hence we must resort to using spectrometric methods of temperature measurement, and the temperatures mentioned above simply refer to the particular method that was used to measure it, and because they do not agree it is one indication that the ICP is not in thermal equilibrium.

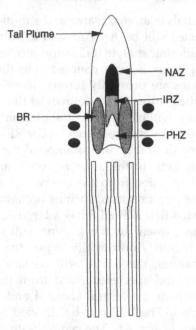

**Figure 4.6** Suggested nomenclature for the various zones in the ICP: pre-heating zone (PHZ); initial radiation zone (IRZ); normal analytical zone (NAZ); boundary region (BR).

**Q.** How is the ICP sustained by radiofrequency energy?

**Q.** Why is a tangential gas flow advantageous?

**Q.** What desirable property does the skin effect impart to the ICP?

**Q.** What are the main species present in the argon ICP?

**Q.** Name the principal axial and radial zones in the ICP – which zone would one normally observe atomic emission?

**Q.** What property (or lack of it) of the ICP makes it difficult to define the temperature?

## 4.4.2  Radiofrequency generators

Two types of design for RF generators have commonly been used, namely crystal-controlled and free-running generators. In a **crystal-controlled** generator the frequency of operation is dictated by an **oscillator circuit** incorporating a crystal oscillating at a fixed frequency. When the power load changes, owing to sample introduction into the plasma, for example, the generator retunes by means of vacuum or airgap **capacitors** to maintain a fixed frequency. **Free-running** generators do away with the necessity for a separate auto-tuning circuit. In this design changes in the power loading are compensated for by slight shifts in the frequency of the oscillation circuit to bring the whole circuit back into resonance. The majority of RF generators operate at 27.12 MHz, although there has been a move towards higher frequencies, particularly 40.68 MHz, which is thought to yield lower continuum background emission and greater stability. Originally RF generators were based on **vacuum tube** technology, although recent instruments have **solid-state** RF generators which are more compact.

---

**Q.** Explain the difference between a free-running and a crystal-controlled radiofrequency generator

---

## 4.4.3 Sample introduction

The commonest form of sample introduction is by means of an aerosol generated using a **pneumatic nebulizer**. Several types of nebulizer can be used. All-glass **concentric** nebulizers (Fig. 4.7a) operate in a similar manner to those used for FAAS. **Cross-flow** nebulizers (Fig. 4.7b) operate by directing a high-velocity stream of gas across the mouth of a capillary

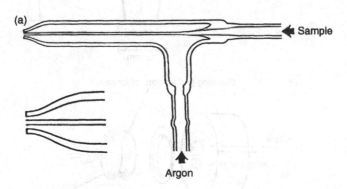

**Figure 4.7** Nebulizers used for ICP-AES: (a) concentric; (b) cross-flow; (c) glass frit (reproduced with permission from Caruso *et al.*, *Spectrochim. Acta*, 1985, **40B**, 3); (d) Hildebrand grid (expanded view) (reproduced with permission from Caruso *et al.*, *J. Anal. At. Spectrom.*, 1987, **2**, 389); (e) V-groove or 'Babington type'.

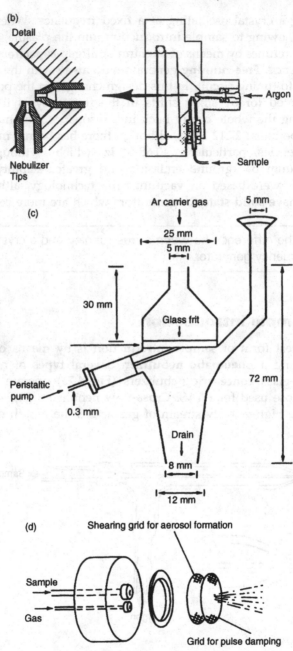

**(b)**

Detail

Argon

Nebulizer
Tips

Sample

**(c)**

Ar carrier gas

5 mm

25 mm

5 mm

30 mm

Glass frit

72 mm

Peristaltic
pump

0.3 mm

Drain

8 mm

12 mm

**(d)**

Shearing grid for aerosol formation

Sample

Gas

Grid for pulse damping

Hildebrand grid

**Figure 4.7** (continued)

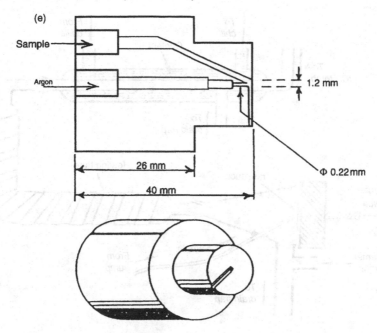

(e)

Sample

Argon

1.2 mm

26 mm

40 mm

Φ 0.22 mm

**Figure 4.7** (continued)

which is immersed in the solution to be nebulized. The drop in pressure causes the solution to be drawn up the tube and shattered into droplets. The main disadvantage of these nebulizers is that they can only tolerate solutions containing less than **0.1–1%** dissolved solids.

The solution to be nebulized is usually pumped to the nebulizer using a **peristaltic pump**, unlike for FAAS, where the solution uptake is by **free aspiration**. The solution is pumped through polymeric tubing [usually poly(vinyl chloride)] and also connecting tubing (usually Teflon) to the nebulizer. Both of these materials can be manufactured to a high degree of purity, hence contamination is minimized. The solution is pumped at a rate of 1–2 ml $min^{-1}$, which is much slower than the 5–10 ml $min^{-1}$ uptake rate for FAAS. This tends to favour the formation of fewer but smaller droplets, which results in less noise but a lower overall sample transport efficiency.

**Glass frit** (Fig. 4.7c) and **grid-type** (Fig. 4.7d) nebulizers operate by running the sample over a glass frit or metal grid, respectively. The nebulizer gas passes through the frit or grid at high velocity, shearing the sample solution into fine droplets. Such nebulizers have greater transport efficiency than pneumatic nebulizers owing to the very fine droplets produced, although they also suffer from salt deposition if solutions containing high dissolved solids are aspirated.

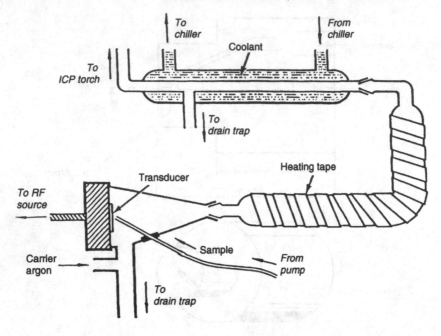

**Figure 4.8**  Schematic diagram of an ultrasonic nebulizer used for ICP-AES (reproduced with permission from the Perkin Elmer Corporation).

**Ultrasonic** nebulizers (Fig. 4.8) operate by directing the sample solution over the surface of a **transducer plate**, usually a **piezoelectric membrane**, which is made to **vibrate** at high frequency by application of RF energy at ca **1 MHz**. A transducer simply converts one form of energy into another, in this case electrical energy into physical movement. As the solution passes over the rapidly vibrating transducer it is literally shattered into very fine droplets, which are swept along in a gas stream to the ICP. Such systems can increase sample transport to 10–15%. However, there is also the requirement to **desolvate** the aerosol to some extent using a heater/condenser, in order to prevent the large amount of solvent vapour which is produced from disturbing the operation of the plasma.

When solutions containing a high concentration of dissolved solids are to be analysed, a **Babington-type** nebulizer (Fig. 4.7e) can be used. In the most common design the sample is pumped through a large sample orifice and runs down a **V-groove** over a pin-hole gas orifice, where the high velocity of the emergent gas shatters it into droplets.

Aerosols generated by nebulization are directed through a **spray chamber**, which is usually constructed from glass, or an inert polymer (Fig. 4.9) The spray chamber prevents larger aerosol droplets from reaching the plasma, as

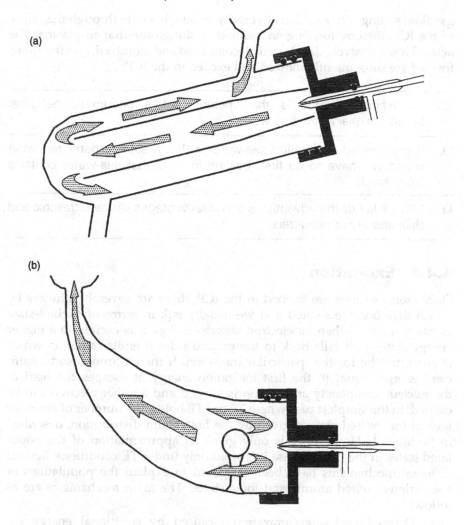

**Figure 4.9** Spray chambers commonly used for ICP-AES: (a), double pass; (b), single pass with impact bead.

they would cause flicker and consequent imprecision, and generally the bulk of droplets are of the order of 2–5 µm in diameter when the aerosol exits the spray chamber. One consequence of this is that the sample transport system is **inefficient**, of the order of 1–2%. On exiting the spray chamber, the gas stream containing the aerosol is directed through the injector tube which forms part of the quartz torch, and thence into the base of the ICP. The injector tube has an internal diameter of ca 2 mm, so the 0.6–1.0 1 min⁻¹ of gas flow exiting it has sufficient velocity to punch a hole through the centre of the ICP, thereby forming an **annular** or **doughnut-shaped** plasma. The aerosol is successively desolvated, decomposed and atomized, and the atoms formed are subsequently ionized and excited in the ICP.

---

**Q.** On which principle is the operation of a pneumatic nebulizer based – explain how this works?

---

**Q.** If you wished to nebulize sea-water with a Meinhard nebulizer, what would you have to do first, bearing in mind that sea-water contains 3.5% NaCl?

---

**Q.** Make a list of the advantages and disadvantages of the ultrasonic and Babington-type nebulizers.

---

## 4.4.4 Excitation

Once atoms or ions are formed in the ICP, there are several pathways by which they become excited and we usually talk in terms of **excited-state** atoms and ions. When an electron absorbs energy it is excited to a higher energy state; as it falls back to the ground state it emits radiation which is characteristic for that particular transition. If the electron absorbs suffi-cient energy, equal to the first ionization energy, it escapes the hold of the nucleus completely and an ion is formed and another electron can be excited. In the simplest case where TE or LTE holds, the number of atoms or ions in the excited state is given by the Boltzmann distribution, described in Section 4.1. However, this only gives an approximation of the popu-lated states in the ICP because it is valid only under TE conditions. Several different mechanisms have been proposed to explain the populations of the various excited atomic and ionic states. The main mechanisms are as follows.

(i) **Thermal** excitation/ionization — caused by collisional energy ex-change between atoms, ions and electrons, e.g.

$$X + e^-_{fast} \longrightarrow X^* + e^-_{slow}$$

$$M + Ar_{fast} \longrightarrow M^* + Ar_{slow}$$

where the asterisk denotes an excited state.

(ii) **Penning** ionization/excitation — caused by collisions between ground-state atoms and argon metastable species, e.g.

$$Ar^m + X \longrightarrow Ar + X^+ + e^-$$

$$Ar^m + X \longrightarrow Ar + X^{+(*)} + e^-$$

(iii) **Charge-transfer** ionization/excitation — caused by the transfer of charge between ions and atoms, e.g.

$$Ar^+ + X \longrightarrow Ar + X^+$$

$$Ar^+ + X \longrightarrow Ar + X^{+(*)}$$

The relative concentrations of excited- and ground-state species are given in Table 4.1.

Most elements are almost completely singly ionized in the argon ICP (a fact which also makes it an ideal ion source for mass spectrometry), hence the majority of the most sensitive emission lines result from atomic transition of ionised species, so-called **ion lines**, with fewer sensitive **atom lines**. Ion lines are usually quoted as, e.g., Mn II 257.610 nm and atom lines as, e.g., Cu I 324.754 nm, with the roman numerals II and I denoting ionic and atomic species, respectively.

## 4.4.5 Monochromators

The commonest **configuration** for an ICP-AES instrument is for the monochromator and detector to view the plasma **side-on** as shown in Fig. 4.10a. This means that there is an optimum **viewing height** in the plasma which yields the maximum signal intensity, lowest background and least interferences, and it is common practice to **optimize** this parameter to obtain the best performance from the instrument. Criteria of merit commonly used include the **signal-to-noise ratio** (SNR), **signal-to-background ratio** (SBR) and net signal intensity. Different elements, and different analytical lines of the same element, will have different optimum viewing heights so a set of **compromise conditions** is usually determined which give satisfactory performance for a selected suite of analytical lines. Emission lines of ionic species, i.e. **ion lines**, tend to emit most strongly in the NAZ, whereas emission lines of atomic species, i.e. **atom lines**, emit most strongly lower down.

An alternative configuration which is gaining popularity is **end-on** viewing of the plasma shown in Fig. 4.10b. This has the advantage that an optimum viewing height does not have to be selected because the analytical signal from the whole length of the central channel is integrated. However, any interferences present are also summed, so that interference effects are expected to be more severe with this configuration.

Atomic emission from the plasma is focused on to the entrance slit of the **monochromator** using a combination of convex or plano-convex lenses or a concave mirror. The combination of focusing optics, monochromator and detector is generally referred to as a spectrometer, although the heart of the device is the monochromator. A monochromator is an instrument that

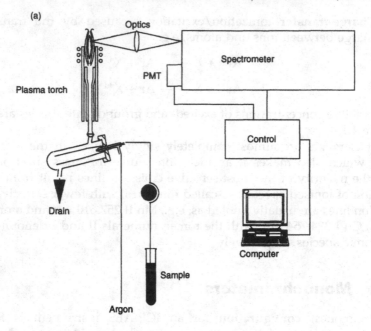

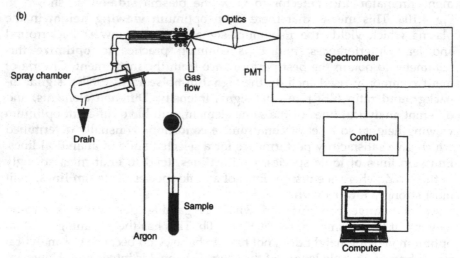

**Figure 4.10** Typical configuration for ICP-AES instruments: (a) side-on viewing of the ICP; (b) axial viewing of the ICP.

can isolate a narrow range of wavelengths (e.g. 1–0.01 nm) anywhere in a comparatively wide spectral range (for atomic absorption spectrometry typically 190–900 nm). Most modern instruments use a **diffraction grating**, which is a mirror having between 600 to 4200 lines per millimetre etched into it. The commonest type of diffraction gratings in use are **reflection gratings**, which are manufactured by depositing a layer of aluminium on the grating blank, then ruling the lines with a diamond tool, to form a **master grating** from which replica gratings can be produced. A more modern method is used to produce **holographic gratings**, obtained by coating the blank with a photosensitive material on to which is projected the interference pattern of two coherent lasers to form the lines. The glass is etched and a layer of aluminium deposited to form the reflective surface. Holographic gratings are less prone to **ghosting** because the line pattern is more uniform, although they are less efficient.

A detailed discussion of the theory of a diffraction grating is beyond the scope of this book. However, a simplified discussion of the mode of operation of a **transmission grating** will illustrate the general principle which can be applied to a reflection grating. A transmission grating is a transparent plate having 2000 lines mm$^{-1}$ ruled in it. Light transmitted through the grating behaves like light emanating through a series of regularly spaced narrow slits. From these, the light will emerge as a series of intersecting wavelets like a series of **semi-circles** emanating from **each slit**, i.e. as if each slit were itself a source of radiation. In a short time the wave systems will recombine to form the original wave-front, the so-called **zero-order**. More interestingly, at a **series of angles**, $\theta$ to the grating, there will be constructive interference for light of a given wavelength. The position of these diffraction wavefronts is given by the **grating equation**:

$$n\lambda = d \sin\theta$$

where $\lambda$ is the wavelength, $d$ the space between the rulings on the grating and $n$ an integer (0, 1, 2, 3,...) called the **order**. Thus, a beam of polychromatic radiation is diffracted into a series of spectra symmetrically located on either side of the normal to the grating. It should be noted that for any given $\theta$ there will be several different wavelengths, albeit in different orders.

The problem of overlapping orders is reduced by blazing the grating. Figure 4.11 shows schematically a blazed reflection grating, where the facets are tilted at an angle $\phi$, the blaze angle, from the surface of the grating. A ray incident at an angle $\alpha$ will be reflected from the groove face at an angle $\beta$, such that $\alpha + \phi = \beta - \phi$. Such a grating is known as an echellete grating and is highly efficient in the diffraction of wavelengths for which specular reflection occurs (i.e. most of the radiation is reflected in a particular order determined by the blaze angle, usually the first order). The wavelength at which specular reflection and first order diffraction coincide is called the

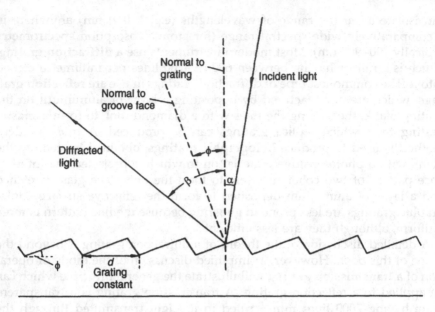

**Figure 4.11**  Schematic diagram of a blazed reflection grating.

blaze wavelength, $\lambda_b$ where

$$\lambda_b = d \sin 2\phi$$

Energy is concentrated in the first order at $\lambda_b$, in the second order at half $\lambda_b$, and so on.

Light striking the surface of the grating will be diffracted to a larger or lesser degree depending on its wavelength, hence a spatial separation of wavelengths is achieved. The diffracted light is **collimated** and **focused**, using mirrors, on to an exit slit, and the diffraction grating can be **scanned** so that different wavelengths are focused on to the slit in turn. A grating may be mounted in a monochromator in several ways. One method, the Ebert mounting (Fig. 4.12a) uses a large spherical mirror to collimate and focus the beam. Czerny and Turner suggested replacing the large, expensive Ebert mirror with two small, spherical mirrors mounted symmetrically, as shown in Fig. 4.12b.

In order to differentiate between wavelengths, a spectrometer with high **resolution** (ca 0.01 nm) is advantageous. Apart from using an echelle grating, there are two ways of achieving this, either by increasing the number of lines per millimetre on the grating or by increasing the focal length of the monochromator. A typical 0.5 or 0.25 m monochromator of

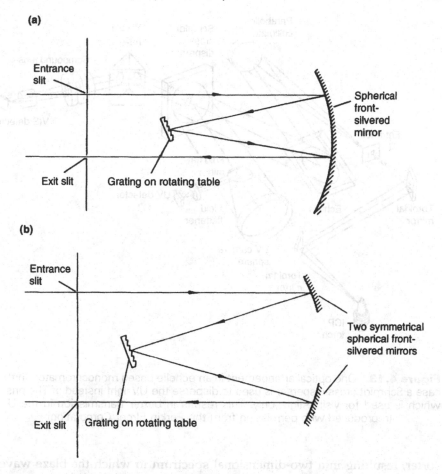

**Figure 4.12** Common grating mountings used in ICP-AES: (a) Ebert; (b) Czerny–Turner.

an atomic absorption instrument does not offer sufficient resolution, so a 0.75 m monochromator is common in modern ICP-AES instruments.

Greater resolution can also be achieved by using an **echelle grating**, which is ruled with step-shaped rulings a few hundred times wider than the average wavelength to be studied (i.e. typically 100 lines per millimetre), in contrast to the general arrangement where the line spacing is similar to the average wavelength. The grooves have large blaze angles, which results in large angles of diffraction. However, the echelle is used at orders of 100 or more, and is thus capable of considerable dispersion. The general arrangement is shown in Fig. 4.13 where the prism, or other dispersing element, is used as an order

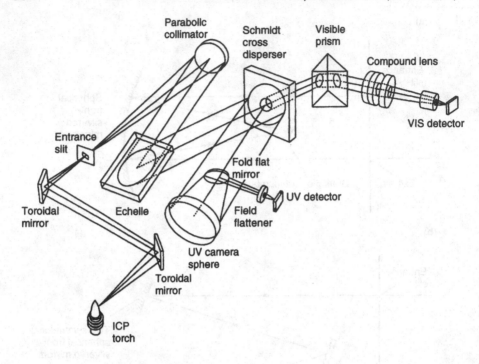

**Figure 4.13**  One optical arrangement in an echelle-based monochromator. In this case a Schmidt cross-disperser is used to disperse the UV light instead of The prism which is used for visible light only. This results in better transmission in the UV (reproduced with permission from the Perkin Elmer Corporation).

sorter, resulting in a two-dimensional spectrum in which the blaze wavelength corresponding to each of the different orders can be used.

**Fourier transform** spectrometers have been developed for the UV–visible region of the spectrum. Such spectrometers make use of a **Michelson interferometer**. The principle advantages of such devices are excellent resolution, the electronic recording of the whole spectrum and extremely fast scanning speed (a few seconds). However, the main disadvantage of this technique stems from the fact that the whole spectrum is collected. Extremely intense emission lines contribute to the noise in all regions of the spectrum, thereby degrading the signal-to-noise ratio of less intense lines.

Other desirable features of a monochromator are **stability** and **multi-element capability**. Initially, direct reading spectrometers, based on a polychromator, were used for simultaneous multi-element analysis, although these were expensive, bulky and generally limited to specific elements. The development of **rapid-scanning** monochromators under computer **control** which **sequentially** tune to analytical wavelengths, greatly increased the flexibility of such systems. More recently, high-resolution echelle monochromators have been used in conjunction with **array detectors** which can be used to monitor much of the spectrum simultaneously.

---

**Q.** What are the main ionization and excitation mechanisms that occur in the ICP?

---

**Q.** What is the short-hand notation for atom and ion lines?

---

**Q.** What desirable properties should a monochromator possess if it is to be used for atomic emission spectrometry?

---

**Q.** Could a coloured filter be used successfully for AES – if not, why not?

---

**Q.** Name two ways of increasing the resolution of a Czerny–Turner monochromator.

---

**Q.** What is the advantage of blazing the grating?

---

**Q.** Why is an echelle monochromator ideal for use with an array detector?

---

## 4.4.6 Detectors

The traditional means of detection has been through the use of a **photomultiplier tube**, which converts the **photon flux** into **electron pulses** which are then amplified. It consists of a partially evacuated transparent envelope containing a photocathode which ejects electrons when struck by electromagnetic radiation. The ejected electrons are accelerated towards a dynode, which then ejects approximately five electrons for every one that strikes. These electrons then strike another dynode, ejecting more electrons and so on until between 9 and 16 dynode stages have been reached. Up to $10^6$ secondary electrons may be ejected from the action of one photon striking the photocathode of a nine-dynode PMT. The electrical current measured at the anode is directly proportional to the radiation reaching the PMT. Figure 4.14 shows schematically the amplification produced by a PMT. The correct choice of photosensitive material used to coat the cathode is important. Usually it is a semiconductive material containing an alkali metal. Different cathode materials have different **response curves**, as can be seen from Fig. 4.15. It can be seen that caesium–antimony cathodes work well up to about 550 nm, but their response at the red end is poor. For these longer wavelengths, a trialkali cathode, sodium–potassium–antimony with a trace

**(a) Side-On Type**

**(b) Head-On Type**

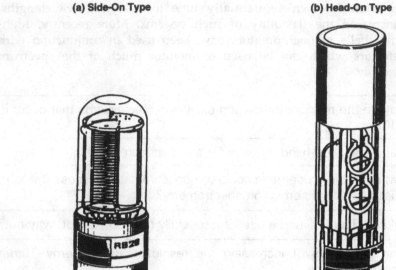

**(c)**

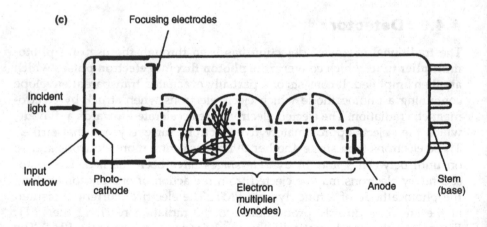

Focusing electrodes

Incident light

Input window

Photo-cathode

Electron multiplier (dynodes)

Anode

Stem (base)

**Figure 4.14** Schematic diagram of photomultiplier tubes: external views of (a) side-on and (b) end-on configurations; (c) mode of operation of the end-on configuration (reproduced with permission from Mamamatsu Photonics).

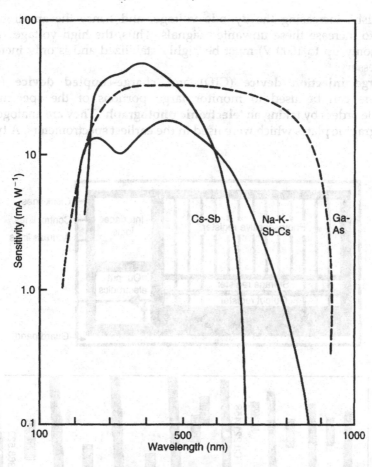

**Figure 4.15** Photomultiplier sensitivity curves.

of caesium, may be preferred. Gallium arsenide cathodes offer a reasonable uniform response over a wide range, up to 850 nm. Response at the low wavelength end (e.g. the arsenic line at 193.7 nm) is often dependent on the material used for the envelope and its transmission at short wavelengths. For certain applications, a photomultiplier which operates only in the ultra-violet region may be preferred. Such a photomultiplier does not respond to daylight and is known as a **solar blind**. Ordinary photomultiplier tubes can be damaged if operated without protection from daylight.

It is important that a photomultiplier gives low noise and a low **dark current**, i.e. low background signal in the absence of photons, usually caused by thermionic emission of photons from the cathode material.

Obviously, increasing the dynode voltage, and hence the amplification, tends to increase these unwanted signals. Thus, the high voltage supply (commonly up to 1000 V) must be highly stabilized and is only increased as necessary.

**Charge injection device** (CID) and **charge-coupled device** (CCD) detectors can be used to monitor large portions of the spectrum in multiple orders by taking an '**electronic photograph**'. They are analogous to photographic plates which were used in the earliest spectrometers. A typical

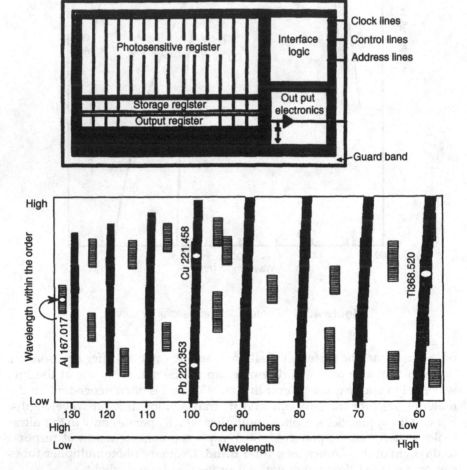

**Figure 4.16** Picture of a CCD detector (top) and schematic diagram of the two-dimensional detector array with respect to wavelength and order (bottom) (reproduced with permission from the Perkin Elmer Corporation).

CCD detector consists of 224 linear photodetector **arrays** on a silicon chip with a surface area of $13 \times 18$ mm (Fig. 4.16). The array segments detect three or four analytical lines of high analytical sensitivity and large dynamic range and which are free from spectral interferences. Each subarray is comprised of pixels. The pixels are photosensitive areas of silicon and are positioned on the detector at $x$-$y$ locations that correspond to the locations of the desired emission lines generated by an echelle spectrometer. The emission lines are detected by means of their location on the chip and more than one line may be measured simultaneously. The detector can then be electronically wiped clean and the next sample analysed. The advantages of such detectors are that they make available as many as ten lines per element, so lines which suffer from interferences can be identified and eliminated from the analysis. Compared with many PMTs, a CCD detector offers an improvement in quantum efficiency and a lower dark current.

---

**Q.**  How is amplification achieved in a photomultiplier tube?

---

**Q.**  What type of photomultiplier would you choose to determine arsenic at the 197 nm line?

---

**Q.**  What is dark current?

---

**Q.**  Why are CCD detectors used widely in multi-element analysis?

---

## 4.4.7  Data handling

Effective signal handling has become an increasingly important feature of ICP-AES instruments owing to the large amount of data produced. The signal output from the detector is usually amplified, converted into a digital signal with an **analogue-to-digital converter** (ADC) and input into a **microcomputer**. This facilitates rapid and efficient handling of the data, and performs such tasks as logging of samples, analytical wavelengths and plasma operating conditions, control of an autosampler for unattended operation, calibration and calculation of the results and report writing. Long gone are the days of chart recorders and analogue readout meters, although these still have their place in the research laboratory — a fact often ignored by the manufacturers of modern instruments.

## 4.4.8  Performance characteristics

ICP-AES is characterized by **low detection limits** of the order of $1$–$100$ ng ml$^{-1}$, as shown in Fig. 4.17, because of its high excitation temperature compared with the flame. It has a **large linear dynamic range**

| 1 | 2 | 3 | 4 | 5 | 6 | 7 | 8 | 9 | 10 | 11 | 12 | 13 | 14 | 15 | 16 | 17 | 18 |
|---|---|---|---|---|---|---|---|---|---|---|---|---|---|---|---|---|---|
| H | | | | | | | | | | | | | | | | | He |
| Li 2 | Be 0.1 | | | | | | | | | | | B 3 | C 40 | N | O | F | Ne |
| Na 5 | Mg 0.1 | | | | | | | | | | | Al 5 | Si 4 | P 50 | S 50 | Cl | Ar |
| K 70 | Ca 0.1 | Sc 0.3 | Ti 0.5 | V 3 | Cr 3 | Mn 0.4 | Fe 3 | Co 3 | Ni 7 | Cu 1 | Zn 3 | Ga 10 | Ge 20 | As 20 | Se 50 | Br | Kr |
| Rb 100 | Sr 0.06 | Y 0.3 | Zr 0.8 | Nb 5 | Mo 5 | Tc | Ru 8 | Rh 8 | Pd 2 | Ag 2 | Cd 2 | In 40 | Sn 30 | Sb 40 | Te 50 | I | Xe |
| Cs 50000 | Ba 0.1 | La 1 | Hf 15 | Ta 20 | W 20 | Re 10 | Os 0.4 | Ir 20 | Pt 30 | Au 5 | Hg 50 | Tl 50 | Pb 40 | Bi 20 | Po | At | Rn |
| Fr | Ra | Ac | | | | | | | | | | | | | | | |

Lanthanides:

| Ce 10 | Pr 40 | Nd 50 | Pm | Sm 40 | Eu 3 | Gd 20 | Tb 20 | Dy 10 | Ho 6 | Er 10 | Tm 5 | Yb 2 | Lu 1 |
|---|---|---|---|---|---|---|---|---|---|---|---|---|---|

Actinides:

| Th 70 | Pa | U 20 | Np | Pu | Am | Cm | Bk | Cf | Es | Fm | Md | No | Lw |
|---|---|---|---|---|---|---|---|---|---|---|---|---|---|---|

**Figure 4.17** Typical detection limits for ICP-AES in ng ml$^{-1}$

(up to six orders of magnitude) because of its optical thinness. Because of the high temperature most samples are completely atomized and the technique suffers from few **chemical interferences** compared with the flame. However, the nature of the sample introduction system means that it does suffer from **sample transport effects**. Such effects are caused by the sample and standard solutions having different physical characteristics such as viscosity, surface tension and volatility. This can result in different nebulization and sample transport efficiencies, and manifest itself as apparent **suppression** or **enhancement** of the analytical signal for the sample compared with the standard. Another type of interference is caused by a large excess of **easily ionizable elements** (EIEs), usually the alkali or alkaline earth elements, in the sample matrix. Unless the plasma operating conditions are carefully optimized a suppression in the analytical signal can result. It is thought that the EIEs cause a reduction in the excitation temperature, or cause **ambipolar diffusion** of the analyte out of the central channel. These types of interference can be overcome by **matrix matching** the samples and standards, i.e. ensuring that the matrix in which the standard is prepared is the same as that of the sample. This is possible for samples such as sea-water, blood and urine, or a well characterized sample which is analysed on a routine basis, although for the majority of samples the matrix will be unknown, in which case the method of **standard additions** can be used.

It is also possible to use an **internal standard** to correct for sample transport effects, instrumental drift and short-term noise, if a simultaneous multi-element detector is used. Simultaneous detection is necessary because the analyte and internal standard signals must be **in-phase** for effective correction. If a sequential instrument is used there will be a time lag between acquisition of the analyte signal and the internal standard signal, during which time short-term fluctuations in the signals will render the correction inaccurate, and could even lead to a degradation in precision. The element used as the internal standard should have similar chemical behaviour as the analyte of interest and the emission line should have similar excitation energy and should be the same species, i.e. ion or atom line, as the analyte emission line.

**Spectroscopic interferences** can also manifest themselves, either as an increase in the **continuum background emission** or as **line overlap**, especially if samples with a complex matrix, or organic solvents, are analysed. An increase in the continuum background emission can be easily compensated for by subtraction of the background adjacent to the analytical line. For a sloping background then measurements must be made on both sides of the line and usually the mean value is subtracted. These options are summarized in Fig. 4.18. Line overlap is a particular problem when an element, present in large excess in the matrix, has an emission line close to,

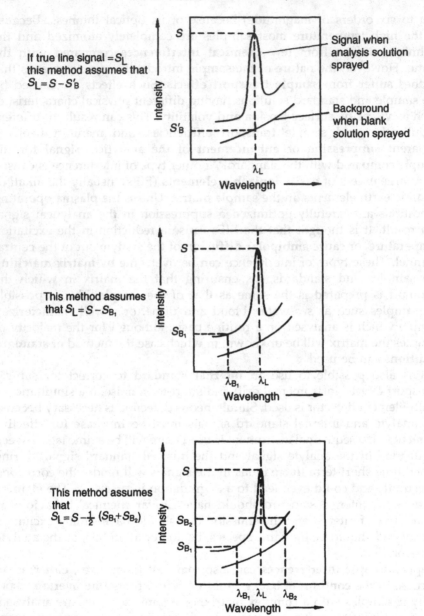

**Figure 4.18** Background correction methods used in atomic emission spectrometry.

or coincident with, the analytical line. When the overlap is not too severe, background correction can be performed by plotting the real background from the spectral information. When complete overlap occurs, it is usually best to utilize an alternative line.

---

**Q.** What are the main types of spectroscopic and non-spectroscopic interferences?

---

**Q.** Why must care be exercised when using internal standardization?

---

**Q.** How would you correct for an increase in the continuum background emission caused by aluminium?

---

## 4.4.9 Applications

ICP-AES has been used to analyse a wide variety of sample types. It is usually necessary to present the sample in the form of a **liquid**, although solids can be analysed directly in the form of **slurries**, by **direct insertion**, **laser ablation** or **electrothermal vaporization** (see Chapter 7).

Methods that have been used to prepare samples for ICP-AES are similar to those used for flame spectroscopy, including sample **digestion** with acids, **fusions**, **solvation** in organic solvents or **thermal decomposition** and subsequent **dissolution**. Because the ICP has such a high temperature it is possible to introduce solid samples while still achieving nearly complete atomization of the sample. This makes it ideal for the analysis of slurried samples of refractory compounds, minerals and coals, which are difficult to dissolve without the use of perchloric and hydrofluoric acid, or fusion methods. Biological materials can generally be digested in one or a combination of nitric, sulphuric, hydrochloric and phosphoric acid. Polymeric materials usually require prior thermal decomposition in a muffle furnace or attack with a strongly oxidizing acid such as sulphuric acid. Monomers and oils can be dissolved in an appropriate organic solvent such as methyl isobutyl ketone (MIBK) or xylene. One limitation of the ICP is that it cannot readily accept volatile organic solvents. Unlike a chemical flame which derives its energy from chemical oxidation of the fuel, the ICP is sustained by coupling electromagnetic energy through a radiofrequency circuit, which sustains the collisional ionization of the support gas. If volatile organic solvent is introduced into the plasma, the efficiency of the RF coupling will be disturbed and an **impedance mis-match** will occur, causing the RF generator to operate with high reflected power, causing damage to the generator. Modern instruments have RF matching networks or free-running generators which can accommodate this; however, if the solvent loading is very high then the energy needed for dissociation of the organic solvent will be too high for the plasma to be sustained.

## 4.5   OTHER FLAME-LIKE PLASMA SOURCES

### 4.5.1   Microwave plasmas

The **capacitatively coupled microwave plasma** is formed by coupling a **2450 MHz magnetron**, via a **coaxial waveguide**, to metal plates or a torch where the plasma is formed. Considerable problems have been encountered with this low-cost plasma, particularly from easily ionizable elements which cause dramatic changes in the excitation temperature in the plasma.

Alternatively, a **microwave-induced plasma** (MIP) may be formed in a **resonant cavity**, using a similar generator. Powers up to 1.5 kW may be achieved, although a maximum of 200 W is more common. If a small flow of argon (e.g. 300 ml min$^{-1}$) is passed through a small bore (2 mm internal diameter) quartz tube placed in the cavity, and seeded with electrons, a self-sustaining plasma will form. Several types of cavity have been used, including the **3/4-wave Broida** cavity and the **1/4-wave Evenson** cavity, but the most popular configuration has been the **TM$_{010}$ Beenakker cavity** (Fig. 4.19) because it is capable of sustaining a **helium plasma** at atmospheric pressure. More recently, the **surfatron** has gained popularity,

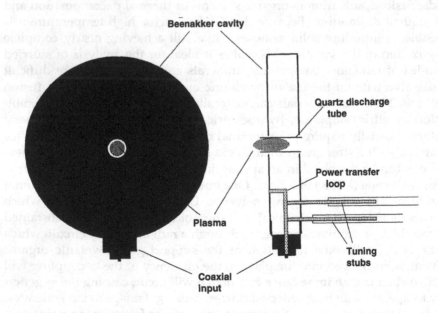

**Figure 4.19**   Schematic diagram of a Beenakker TM$_{010}$ microwave cavity.

in which the MIP is launched using a surface wave. The temperature of the helium MIP is difficult to define as it is not in LTE. The excitation temperature is in the region of 5000–7000 K, but the neutral gas temperature is between 1000 and 2000 K. However, the presence of high-energy electrons and metastable excited-state species means that the MIP is a highly efficient excitation source. If helium is used as the plasma gas, the existence of high-energy helium metastable species means that elements such as chlorine, fluorine, nitrogen and oxygen are excited efficiently.

Although inexpensive and compact, the microwave plasma suffers from low tolerance to solution samples and from chemical interferences. The latter arise because of the low gas temperature. These problems can be overcome to some extent by using a high-power MIP (up to 1.5 kW), although this increases the complexity of the generator and tuning network, and requires the use of a high gas-flow tangential torch, much like the ICP. The MIP has found its greatest application as an **element-selective detector** for **gas chromatography**. The advantages of element-selective detection are that compounds which elute from a gas chromatograph can be identified **unequivocally** with respect to their constituent atoms, and not just on the basis of their retention time. **Organometallic species** containing tin, mercury and lead can be determined down to picogram levels. Other applications include the determination of phosphorus and sulphur in oils and halogenated pesticides. These advantages have been exploited in commercial instruments, the most recent of which incorporates a diode-array detector for the simultaneous monitoring of up to four elements (Fig. 4.20).

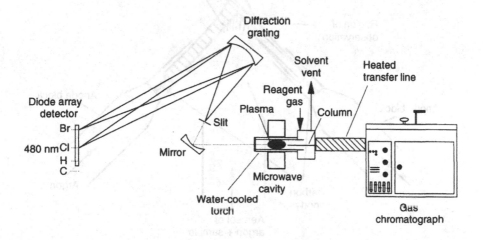

**Figure 4.20** Schematic diagram of a gas chromatography atomic emission detection (GC-AED) instrument.

## 4.5.2 Direct current plasmas

A **DC arc plasma** can be generated between two electrodes and a portion of the discharge can be moved out of the primary arc column using gas flows to bend the arc. One commercial variant of this employs three electrodes (two anodes and a cathode) to form an **inverted 'Y' discharge** (Fig. 4.21). This plasma accepts liquid samples in aqueous or organic solvents. It is unlikely that the sample actually penetrates the highest temperature part of the discharge (7000–9000 K). In any case, the high plasma background does not permit observation in this region, which is made instead in

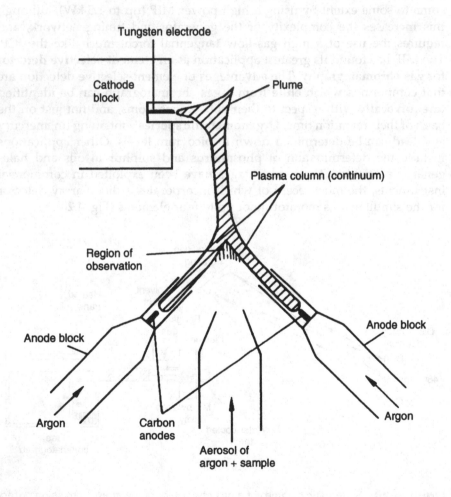

**Figure 4.21** Schematic diagram of the three-electrode direct current plasma.

the angle of the 'Y'. The excitation temperature is about 5500 K. The DC plasma is comparatively **economical** to run, consuming less than 1000 W and about 8 l min$^{-1}$ of argon. Unfortunately there may be a **tungsten background** spectrum from the electrodes. A very troublesome problem with this plasma appears to be the substantial enhancement of emission caused by easily ionizable elements. **Buffering** with lithium or barium to overcome the problem has been proposed. An echelle monochromator (or polychromator if the exit slit is replaced with a cassette of about 20 elemental slits) has been used extensively with this type of plasma, offering excellent resolution, precision and accuracy, with lower detection limits than with flame emission, although not as good as with ICP-AES.

---

**Q.**   What are the main differences between the MIP and ICP?

---

**Q.**   What interference problems may be encountered with microwave plasmas?

---

**Q.**   Why are microwave plasmas used widely as detectors for gas chromatography?

---

**Q.**   Where is the main analytical observation zone in the DCP?

---

**Q.**   What are the advantages and disadvantages of the DCP?

---

## 4.6   SOLID SAMPLING PLASMA SOURCES

### 4.6.1   Arcs and sparks

An **arc** is a continuous electrical discharge of high energy between two electrodes. In the **DC arc**, the sample is usually packed into an **anode electrode,** or the sample itself becomes the anode as, for example, in metal analysis. The arc is struck and the sample is vaporized into the discharge region, where the excitation and emission occur. Because relatively large amounts of sample are excited, the detection limits are low, but the **unstable** nature of the discharge leads to poor precision. Additionally, the intensity of emission is highly dependent on the matrix. If the light from the discharge is dispersed by a prism, and a camera of long focal length (e.g. 1 m) is used, a photograph of the spectrum gives good resolution and can be used for the rapid **qualitative** identification of the elements in an unknown sample.

If an **intermittent discharge** is used (e.g. an **AC spark**), the precision can be greatly improved. However, the detectability is poorer because only relatively small amounts of sample will be vaporized. Modern electrical discharges combine the characteristics of both arc and spark (e.g. a unidirectional spark may be used so that the intermittent discharge always excites the sample as the cathode) to obtain optimal detectability and precision for quantitative work. Further improvements have been made by **sheathing** the electrodes with argon to reduce the amount of self-absorption by the cooled sample, reduce the background and stabilize the discharge. Attempts to reduce the influence of surface effects on the final result by careful pre-sparking are also made. Such discharges are used very widely in polychromator systems, with computer control of the data acquisition and calculation of the several corrections needed because of inter-element effects and spectral interferences. These instruments are referred to as **direct reading** spectrometers and, in conjunction with arc/spark sources, are widely used in the **steel industry** because they can be used for the rapid analysis of solid samples. Mobile spectrometers incorporating arc and spark discharges have recently become available and are used extensively for spot testing raw materials and scrap.

## 4.6.2   Glow discharges

**Glow discharges** can be used to analyse both conducting and non-conducting samples. A glow discharge is formed between two electrodes in an inert gas (e.g. argon) at **low pressure** (0.1–10 Torr). The sample forms one of the electrodes, usually the cathode with the wall of the discharge chamber forming the anode (Fig. 4.22). With **DC glow discharges**, non-conducting samples must be mixed with a conducting material (e.g. graphite) and pressed into a pellet, while **radiofrequency glow discharges** allow the direct analysis of **non-conducting samples**. When the discharge is initiated, argon atoms are accelerated across the **dark space** towards the sample surface where they dislodge several atoms in a process known as **sputtering**. The sputtered atoms are then ionised and excited in the **negative glow** region of the discharge.

One popular configuration is the **Grimm source**, which accepts samples in the form of discs. Such sources usually operate at 500–1000 V, 25–100 mA and 1–5 Torr, with detection limits of approximately 0.1 ppm. Another configuration is the **hollow-cathode lamp** in which the sample can be either machined as a hollow cathode, evaporated to dryness (if a solution) or pressed (if a powder) into a hollow cathode made of pure graphite. Typical operating conditions are 200–500 V, 10–100 mA, and 0.1–1.0 Torr, with detection limits in the range 0.1–10 ppm.

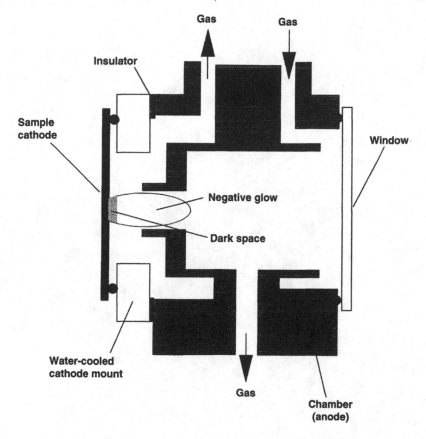

**Figure 4.22** Schematic diagram of a glow discharge cell.

**Q.** Why are arc/spark emission spectrometers ideal for industrial applications?

**Q.** What property must the sample possess in order to be amenable to analysis by DC glow discharge spectrometry?

**Q.** Where in the glow discharge is the most probable place to observe atomic emission?

# 5 INDUCTIVELY COUPLED PLASMA MASS SPECTROMETRY

Inductively coupled plasma mass spectrometry (ICP-MS) is the marriage of two well established techniques, namely the **inductively coupled plasma** and **mass spectrometry**. The ICP has been described as an ideal ion source for inorganic mass spectrometry. The **high temperature** of the ICP ensures almost **complete decomposition** of the sample into its constituent atoms, and the ionization conditions within the ICP result in highly **efficient ionization** of most elements in the Periodic Table and, importantly, these ions are almost exclusively **singly charged**.

A schematic diagram of an ICP-MS instrument is shown in Fig. 5.1. The 'ICP part' bears an almost exact resemblance to the ICP used for atomic emission spectrometry, with the obvious exception that it is turned on one side. Indeed, sample introduction systems, radiofrequency generators and the nature of ICP itself are often the same for ICP-MS and ICP-AES systems, with the usual variations between individual manufacturers.

## 5.1  SAMPLE INTRODUCTION

The sample is introduced into the ICP as a **liquid** which must usually contain less than **0.1% dissolved solids** to prevent salt build-up on the **nickel** cones (see Section 5.3). This is in contrast to ICP-AES, which can tolerate up to 1% dissolved solids. The sample is converted to an **aerosol** by means of a **pneumatic nebulizer**, and the droplets pass through a **spray chamber**, into the injector tube of the **quartz torch** and thence into the **central channel** of the ICP. These processes are identical with those described for ICP-AES (see Section 4.4.3), and the different types of

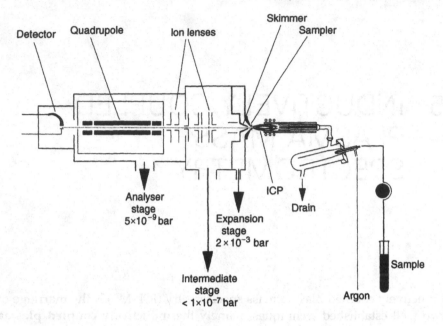

**Figure 5.1**  Schematic diagram of a commercial inductively coupled plasma mass spectrometer.

nebulizer, spray chamber and torches used for ICP-AES can also be used successfully with ICP-MS. Hence, liquid sample introduction systems for ICP-MS have between only **1 and 15% efficiency**, in common with those used for ICP-AES.

Once in the ICP, the aerosol droplets are successively **desolvated, decomposed** and **ionized** by the high temperature of the plasma, and it is at this point that the techniques of ICP-AES and ICP-MS diverge.

## 5.2  IONIZATION

If an electron absorbs sufficient energy, equal to its **first ionization energy**, it escapes the atomic nucleus and an ion is formed. In the ICP the major mechanism by which ionization occurs is **thermal ionization**. When a system is **in thermal equilibrium**, the **degree of ionization** of an atom is given by the **Saha equation**:

$$\frac{n_i n_e}{n_a} = 2\frac{Z_i}{Z_a}\left(2\pi m k \frac{T}{h^2}\right)^{3/2} \exp(-E_i/kT) \qquad (5.1)$$

where $n_i$, $n_e$ and $n_a$ are the number densities of the ions, free electrons and atoms, respectively, $Z_i$ and $Z_a$ are the ionic and atomic partition functions, respectively, $m$ is the electron mass, $k$ is the Boltzmann constant, $T$ is the temperature; $h$ is Planck's constant and $E_i$ is the first ionization energy. In this case, ionization is effected by **ion–atom** and **atom–atom collisions**, where the energy required for ionization is derived from **thermal agitation** of the particles. The degree of ionization is dependent on the **electron number density**, the **temperature** and the **ionization energy** of the element in question. Taking the average electron number density for an argon ICP to be $4 \times 10^{15}$ cm$^{-3}$ and the ionization temperature to be 8730 K, then the degree of ionization as a function of first ionization energy, predicted by the Saha equation, is as shown in Fig. 5.2. Most of the elements in the Periodic Table have first ionization energies of less than 9 eV and are over 80% ionized in the ICP. The remaining third are ionized to a lesser extent depending on their first ionization energy, with the most poorly ionized elements being He, Ne, F, O, N < 1% ionized, Kr, Cl 1% to 10%, C, Br, Xe, S 10% to 30% and P, I, Hg, As, Au, Pt 30% to 80%. Such thermal ionization is probably the dominant mechanism of ionization in the ICP.

**Electron impact** ionization can occur when the **kinetic energy** of an incident **free electron** exceeds the ionization energy of the atom so that **collision** may result in **ionization** of the atom and, hence, liberation of a second

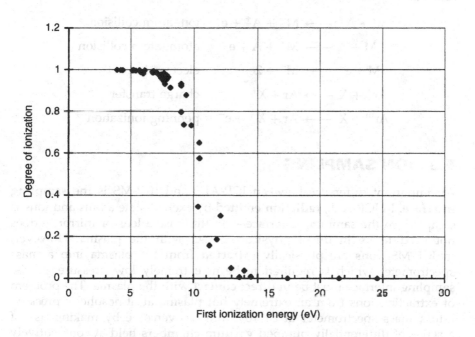

**Figure 5.2** Degree of ionization as a function of first ionization energy calculated using the Saha equation.

free electron. In practice, however, a quantum mechanical treatment reveals that the ionization cross-section is very small when the kinetic energy of the electron is equal to the ionization energy of the atom, but increases up to a maximum at about 100 eV, the rate of ionization being directly proportional to the electron current.

**Charge transfer** occurs when interactions between atoms and ions allow insufficient time for the electrons to adjust themselves to the new conditions, and the charge is transferred from the ion to the neutral atom. This is a complex process which requires that the ionization energies of the two species are similar.

Some atomic species can undergo excitation to excited states which are relatively **long lived**, so-called **metastable states**, and can therefore undergo several collisions before de-excitation. If the metastable atom collides with a neutral molecule it can induce ionization of the neutral molecule with consequent de-excitation of the metastable atom, a process called **Penning ionization**. The likelihood of this occurring is greatest when the potential energy of the metastable species is similar to the ionization energy of the neutral atom.

The various types of ionization processes that occur in the ICP are as follows:

$$M + A^+ \longrightarrow M^+ + A^+ + e^- \quad \text{ion–atom collision}$$

$$M + A \longrightarrow M^+ + A + e^- \quad \text{atom–atom collision}$$

$$M + e^- \longrightarrow M^+ + 2e^- \quad \text{electron impact}$$

$$Ar^+ + X \longrightarrow Ar + X^+ \quad \text{charge transfer}$$

$$Ar^m + X \longrightarrow Ar + X^+ + e^- \quad \text{penning ionization}$$

## 5.3  ION SAMPLING

An important difference between ICP-AES and ICP-MS is the **sampling interface**. In ICP-AES, **radiation** emitted by excited-state atoms and ions is sampled, so the sampling interface — in this case a lens or mirror — does not need to be in direct physical contact with the plasma. However, in ICP-MS, **ions** are physically **extracted** from the plasma into a mass spectrometer which is required to be at extremely low pressure, so the sampling interface must be in **direct contact** with the plasma. The problem of extracting ions from an extremely hot plasma at atmospheric pressure into a mass spectrometer at ca $10^{-9}$ atm is overcome by making use of a series of differentially pumped **vacuum chambers** held at consecutively lower pressures. A schematic diagram of the ICP-MS sampling interface is shown in Fig. 5.3. The ICP is aligned so that the central channel is

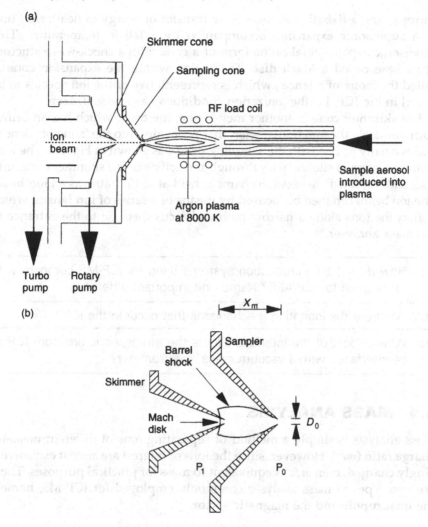

**Figure 5.3** Schematic diagrams of (a) the ICP-MS interface and (b) the supersonic expansion formed in the expansion chamber, showing the barrel shock and position of the Mach disc.

axial with the tip of a **water-cooled sampling cone**, made of nickel or copper, which has an orifice approximately 1 mm in diameter. The pressure behind the sampling cone is reduced, by means of a vacuum pump, to ca $2 \times 10^{-3}$ atm so the plasma gases, together with the analyte ions, **expand** through the sampling orifice to form a **shock-wave** structure as shown in Fig. 5.3. This expansion is **isentropic** (i.e. there is no change in the total entropy) and **adiabatic** (i.e. there is no transfer of energy as heat), resulting in a **supersonic expansion** accompanied by a fall in temperature. This supersonic expansion takes the form of a cone with a shock-wave stucture at its base called a **Mach disc**. The region within the expansion cone is called the '**zone of silence**', which is representative of the ion species to be found in the ICP, i.e. the ionization conditions have been '**frozen**'.

The **skimmer cone** is another metal cone, the tip of which has an orifice approximately 0.7 mm in diameter, that protrudes into the 'zone of silence', and is axially in-line with the sampling orifice as shown in Fig. 5.3. The ions from the 'zone of silence' pass through the orifice in the skimmer cone, into a second intermediate vacuum chamber held at $< 10^{-7}$ atm, as an **ion beam** The ion beam can then be focused by means of a series of **ion lenses**, which deflect the ions along a narrow path and focus them on to the entrance to the **mass analyser**.

---

**Q.**  How do sample introduction systems used for ICP-MS compare with those used for ICP-AES? Name one important difference.

---

**Q.**  What are the ionization mechanisms that occur in the ICP?

---

**Q.**  What aspect of the interface allows the atmospheric pressure ICP to be interfaced with a vacuum mass spectrometer?

---

## 5.4  MASS ANALYSIS

Mass analysis is simply a method of separating ions of different **mass-to-charge ratio** ($m/z$). However, since the ions of interest are almost exclusively **singly charged**, then $m/z$ is equivalent to mass for practical purposes. There are two types of mass analyser commonly employed for ICP-MS, namely the **quadrupole** and the **magnetic sector**.

### 5.4.1  Quadrupole mass analysis

Quadrupoles are comprised of **four metal rods**, ideally of **hyperbolic cross section**, arranged as shown in Fig. 5.4. A combination of **radiofrequency** (RF) and **direct current** (DC) voltages are applied to each pair of rods, which creates an **electric field** within the region bounded by the rods. Depending on the **RF/DC ratio**, the electric field between the rods will allow ions in a narrow $m/z$ range to pass, typically 0.8 $m/z$ — just how narrow will depend on a number of factors which influence the resolution. Hence, by changing the RF/DC ratio in a controlled manner, the quadrupole can be

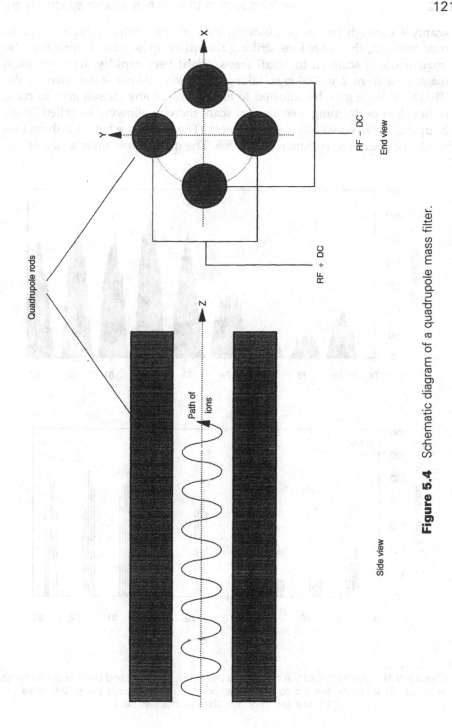

**Figure 5.4** Schematic diagram of a quadrupole mass filter.

**scanned** through the range allowing ions of consecutively higher $m/z$ to pass through, the other ions striking the quadrupole rods. In practice, the quadrupole is scanned in **small steps**, albeit very **rapidly**, with a typical mass scan from 2 to 260 $m/z$ taking less than 100 ms. Alternatively, the RF/DC voltage may be adjusted to allow ions of any chosen $m/z$ to pass, rather than performing a sequential scan, thereby allowing so-called '**peak hopping**' between widely separated $m/z$. The difference between these two modes of operation is shown in Fig. 5.5. The quadrupole mass analyser can

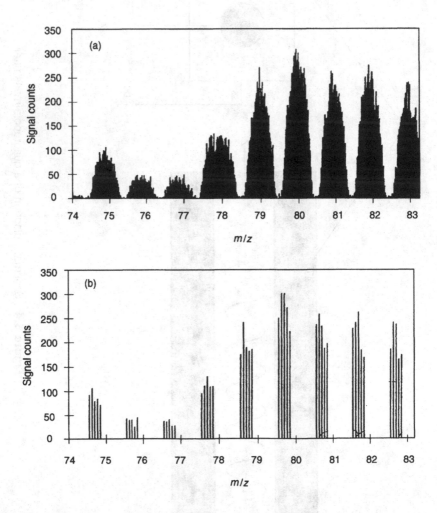

**Figure 5.5**  Illustration of the difference between scanning and peak-hopping mass analysis. In (a) there are up to 100 data points defining each peak, whereas in (b) there are only five data points per peak.

only be operated in **sequential** mode, although the speed with which this can be achieved makes it seem almost like simultaneous mass analysis. The quadrupole mass analyser has the advantage of being cheap, reliable and compact, with single mass resolution which is sufficient for most applications, and is therefore the most commonly used mass analyser. However, if an extremely high degree of resolution or true simultaneous mass analysis is required, then a magnetic sector must be used.

## 5.4.2 Magnetic sector mass analysis

Magnetic sector mass analysers rely on the fact that ions are **deflected** by a **magnetic field**, with ions of greater mass or charge being deflected to a greater extent. A typical **double-focusing** instrument, combining both an **electric** and a **magnetic sector** in normal geometry, is shown in Fig. 5.6. The electric sector can be placed either before or after the magnetic sector with the former termed normal and the latter reverse geometry, indicating the chronological order in which the two techniques were developed. In one typical commercial instrument the ions are **acclerated** after they are skimmed from the plasma, then travel through the electric sector, which acts as an energy filter. The ions are then deflected in a **single plane** by the magnetic field, with the degree of deflection increasing with increasing $m/z$. A mass spectrum can be generated by **scanning** the magnetic field, allowing ions of consecutively higher $m/z$ to pass through a **slit**, and then detected. Alternatively, the magnetic and electric field strengths can be held constant and several detectors arranged in an array around the plane in which the

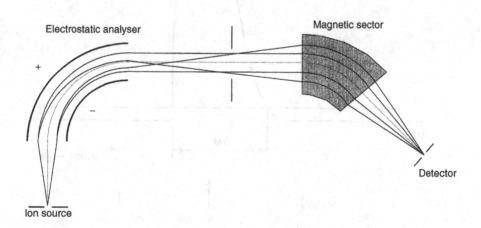

**Figure 5.6** Schematic diagram of a high-resoution mass spectrometer using magnetic and electric sectors.

ions are deflected (cf. polychromator used for optical spectrometry), thereby allowing truly **simultaneous** mass analysis. Magnetic sector mass analysers are more expensive and less easy to operate than quadrupoles, mainly because they require more stable operating environments. Magnetic sectors cannot be scanned as rapidly as quadrupoles, although modern instruments are capable of scanning 2–260 $m/z$ in less than 1 s, and they are also capable of simultaneous operation for a limited number of masses.

### 5.4.3  Resolution

The main advantage of a magnetic sector is the high degree of **resolution** obtainable. Resolution is defined as

$$R = \frac{M}{\Delta M} \tag{5.2}$$

where $R$ is resolution, $M$ is mass (strictly $m/z$) and $\Delta M$ is peak width at 5% peak height. This is concept is illustrated better in Fig. 5.7. The resolution

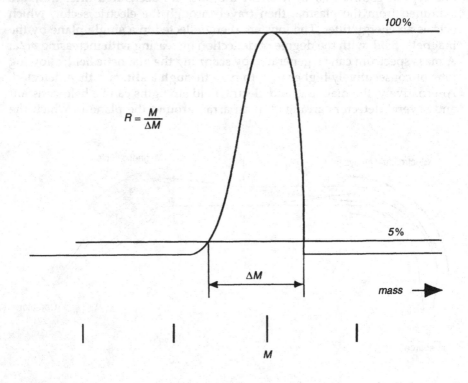

**Figure 5.7**  Definition of mass resolution, $R$.

**Table 5.1** Common spectroscopic interferences caused by molecular ions, and the resolution that would be necessary to separate the analyte and interference peaks in the mass spectrum.

| Analyte ion | | Interfering ion | | |
|---|---|---|---|---|
| Nominal isotope | Accurate $m/z$ | Nominal isotopes | Accurate $m/z$ | Resolution required |
| $^{51}$V | 50.9405 | $^{16}$O$^{35}$Cl | 50.9637 | 2580 |
| $^{56}$Fe | 55.9349 | $^{40}$Ar$^{16}$O | 55.9572 | 2510 |
| $^{63}$Cu | 62.9295 | $^{40}$Ar$^{23}$Na | 62.9521 | 2778 |
| $^{75}$As | 74.9216 | $^{40}$Ar$^{35}$Cl | 74.9312 | 7771 |

obtainable with quadrupoles is limited by the stabiliy and uniformity of the RF/DC field and by the spread in **ion energies** of the ions. Quadrupoles used in ICP-MS are typically operated at resolutions between 12 and 350, depending on $m/z$, which corresponds to peak widths between 0.7 and 0.8. In comparison, magnetic sectors are capable of resolution exceeding 10 000, resulting in peak widths of 0.008 at 80 $m/z$. For most applications the resolution provided by a quadrupole is sufficient; however, for applications when **spectroscopic interferences** cause a major problem, the resolution afforded by a magnetic sector may be desirable. Table 5.1 gives examples of some common spectroscopic interferences that may be encountered (see Section 5.6) and the resolution required to separate the element of interest from the interference. For example, a particular problem is the determination of **arsenic** in a matrix which contains **chloride** (a common component of most biological or environmental samples). Arsenic is monoisotopic (i.e. it only has one isotope) at $m/z$ 75 and the chloride matrix gives rise to an interference at $m/z$ 75 due to $^{40}$Ar$^{35}$Cl$^+$, so an alternative isotope is not available for analysis. A quadrupole has insufficient resolution to separate the two species but a magnetic sector could do so easily, as shown in Table 5.1.

One further advantage of the magnetic sector compared with the quadrupole is that **ion transmission**, and hence **sensitivity**, are much greater for comparable resolution.

## 5.5 ION DETECTION AND SIGNAL HANDLING

Ion detection can be perfomed by a variety of methods, but the commonest by far is the **channel electron multiplier** shown in Fig. 5.8. This consists of a curved glass tube of approximately 1 mm internal diameter with an inner resistive coating and a flared end. The multiplier can be operated in one of two modes. In the **pulse counting mode** — the most sensitive mode

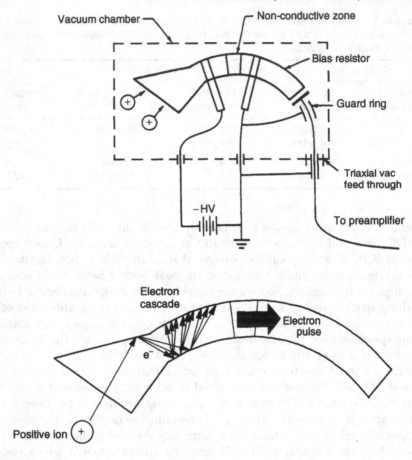

**Figure 5.8** Schematic diagram of an electron multiplier showing and (top) electrical connections for analogue and pulse counting and (bottom) how a single ion gives rise to a large electron pulse.

of operation — a high voltage of between −2600 and −3500 V is applied to the multiplier (Fig. 5.8), which attracts ions into the funnel opening. When a positive ion strikes the inner coating the **collision** results in the ejection of one or more **secondary electrons** from the surface, which are accelerated down the tube by the potential gradient and collide with the wall, resulting in further electron ejection. Hence, an exponential **cascade** of electrons rapidly builds up along the length of the tube, eventually reaching **saturation** towards the end of the tube, resulting in a large **electron pulse** and a consequent **gain** of $10^7$–$10^8$ over the original ion collision. The electron pulses are read at the base of the multiplier and are approximately

50–100 mV and 10 ns in duration. This is almost identical with the way in which a photomultiplier works, the main differences being that instead of **discrete dynodes**, there is one **continuous dynode**, which is the electron multiplier tube itself, and the electron cascade is initiated by an ion instead of a photon. In fact, the electron multiplier is equally sensitive to photons, and because of the intensely bright ICP source, **photon noise** is the major source of background in ICP-MS. Hence the multiplier is positioned off-axis from the ion beam, positive ions being defelected into it by a negatively charged plate opposite the mouth, and the uncharged photons passing by unaffected. Another precaution commonly adopted is the inclusion of a **photon stop** on-axis in the ion lenses, which **shades** the multiplier from the photon-rich source but allows ions to be focused around it.

Alternatively, the multiplier can be operated in the **analogue** mode with a gain of only $10^3$–$10^4$, so that the multiplier does not become saturated and the pulses vary greatly in size. In this mode the applied voltage is between $-500$ and $-1500$ V and the electron pulses are read at the collector electrode (Fig. 5.8) where they are amplified and averaged over a short time interval to allow rapid data acquisition. The greatest sensitivity is achieved with the detector in the pulse counting mode, but the detector will become saturated at counting rates above $10^6$ Hz, which are encountered when the analyte is at a high concentration in the sample. If the detector is switched to the analogue mode it is less sensitive, but can be used for analyte concentrations which are much higher, typically up to three orders of maginitude higher than for pulse counting. Such **dual-mode** operation results in an extremely large **linear dynamic range** of up to nine orders of magnitude but requires that two analytical scans be made to acquire all the data. More recently, electron multipliers with discrete dynodes have come on the market which offer the advantage of **simultaneous** pulse counting and analogue detection.

Another commonly used detector is the **Faraday cup**. This detector is an analogue detector and so has poorer sensitivity than a pulse counting electron multiplier. However, it has the advantage of simplicity (it is essentially only a metal plate used to measure ion current), and it does not suffer from burn-out like an electron multiplier (which must be periodically replaced).

---

**Q.** What are the main differences between the scanning and peak-hopping mode of operation of a quadrupole mass analyser?

---

**Q.** What are the advantages and disadvantages of magnetic sector mass analysers compared with quadrupoles?

---

**Q.** What are the main differences between an electron multiplier operated in pulse counting and analogue mode?

---

## 5.6 PERFORMANCE

Performance characteristics for quadrupole and magnetic sector instruments are shown in Table 5.2. The main advantages that ICP-MS has over other techniques are its **low detection limits**, in the $1-10$ pg ml$^{-1}$ range for quadrupole instruments (Fig. 5.9), large **linear dynamic range** and rapid **multi-element capability**. However, ICP-MS also suffers from a number of interferences.

**Spectroscopic interferences** arise when an interfering species has the same nominal $m/z$ as the analyte of interest. The interfering species can be either an isotope of another element (which are well documented and hence easily accounted for), or a **molecular ion** formed between elements in the sample matrix, plasma gas, water and entrained atmospheric gases. The molecular ions are less easy to correct for since they will vary depending on the nature of the sample matrix. Some common molecular ion interferences are shown in Table 5.3. Many of these interferences can be overcome by choosing an alternative **isotope** of the analyte which is free from interference, although a sacrifice in sensitivity may result. If a 'clean' isotope is not available, then one recourse is to separate the analyte from the matrix before the analysis using chemical extraction and chromatography, or to save up and buy a **magnetic sector** instrument which is capable of resolving the interfering species from the analyte. Many of these interferences are thought to be formed in the interface due to a **secondary discharge**. This discharge can be eliminated by operating the plasma at low power, typically 600 W, and modifying the torch or RF coil. Under these so-called '**cool plasma**' conditions interferences due to ArO$^+$, for example, can be eliminated, although the sensitivity for refractory elements and those with high ionization potentials may be reduced.

**Non-spectroscopic interferences** are caused by the sample matrix, and are manifest as an apparent **enhancement** or **suppression** of the analyte signal in the presence of a concentrated sample matrix. Such effects are thought to be caused primarily by **space charge** in the ion beam, whereby

**Table 5.2** Performance characteristics for ICP-MS.

| Parameter | Quadrupole | Magnetic sector |
|---|---|---|
| Detection limit (pg ml$^{-1}$) | $1-10$ | $0.01-0.1^a$ $1-10^b$ |
| Linear dynamic range | $10^6$ $10^9$ dual mode | $10^6$ $10^{10}$ dual mode |
| Precision (% RSD) | $1-2$ | $1-2$ |
| Mass resolution | 400 | $400-10\,000$ |

$^a$ Resolution 400.
$^b$ Resolution 3500.

**Figure 5.9** Typical detection limits (pg ml$^{-1}$) for quadrupole ICP-MS.

Main table:

| | | | | | | | | | | | | | | | | | |
|---|---|---|---|---|---|---|---|---|---|---|---|---|---|---|---|---|---|
| H 1 | | | | | | | | | | | | | | | | | He |
| Li 1 | Be 1 | | | | | | | | | | | B 50 | C 10000 | N | O | F | Ne |
| Na 50 | Mg 50 | | | | | | | | | | | Al 50 | Si 500 | P 500 | S 10000 | Cl 10000 | Ar |
| K 500 | Ca 500 | Sc 50 | Ti 50 | V 5 | Cr 5 | Mn 0.4 | Fe 50 | Co 1 | Ni 5 | Cu 5 | Zn 5 | Ga 1 | Ge 50 | As 5 | Se 50 | Br 50 | Kr |
| Rb 1 | Sr 1 | Y 5 | Zr 5 | Nb 5 | Mo 5 | Tc 5 | Ru 5 | Rh 5 | Pd 5 | Ag 5 | Cd 5 | In 1 | Sn 5 | Sb 5 | Te 50 | I 50 | Xe |
| Cs 1 | Ba 1 | La 5 | Hf 5 | Ta 5 | W 5 | Re 5 | Os 5 | Ir 5 | | Au 5 | Hg 1 | Tl 1 | Pb 1 | Bi 1 | Po | At | Rn |
| Fr | Ra | Ac | | | | | | | | | | | | | | | |

Lanthanides and actinides:

| | | | | | | | | | | | | | | |
|---|---|---|---|---|---|---|---|---|---|---|---|---|---|---|
| Ce 1 | Pr 1 | Nd 1 | Pm | Sm 1 | Eu 1 | Gd 1 | Tb 1 | Dy 11 | Ho 1 | Er 1 | Tm 1 | Yb 1 | Lu 1 | |
| Th 1 | Pa 1 | U 1 | Np 1 | Pu 1 | Am | Cm | Bk | Cf | Es | Fm | Md | No | Lw | |

**Table 5.3**  Some commonly occurring spectroscopic interferences caused by molecular ions derived from plasma gases, air and water and the sample matrix.

| Molecular ion interference | Analyte ion interfered with | Nominal $m/z$ |
|---|---|---|
| $O_2^+$ | $S^+$ | 32 |
| $N_2^+$ | $Si^+$ | 28 |
| $NO^+$ | $Si^+$ | 30 |
| $NOH^+$ | $P^+$ | 31 |
| $Ar^+$ | $Ca^+$ | 40 |
| $ArH^+$ | $K^+$ | 41 |
| $ClO^+$ | $V^+$ | 51 |
| $CaO^+$ | $Fe^+$ | 56 |
| $ArO^+$ | $Fe^+$ | 56 |
| $ArN^+$ | $Cr^+$, $Fe^+$ | 54 |
| $NaAr^+$ | $Cu^+$ | 63 |
| $Ar_2^+$ | $Se^+$ | 80 |

positive analyte ions are **repelled** from the ion beam by the high positive charge of the matrix ions, with low-mass ions being relatively more affected than high-mass ions. Such interferences are usually compensated for by using an **internal standard** or by separating the matrix from the analyte before analysis. Because ICP-MS has such low detection limits it is also possible to dilute the sample to such an extent that the interference becomes negligible.

---

**Q.**  What are the two main types of interferences encountered in ICP-MS and how do they differ?

---

**Q.**  Which common method could you use to eliminate both non-spectroscopic and many spectroscopic interferences?

---

## 5.7  APPLICATIONS

The applications for ICP-MS are broadly similar to those for ICP-AES, although the better sensitivity of the former has resulted in applications such as the determination of ultra-low levels of impurities in semiconductors and long-lived radionuclides in the environment. Also, ICP-MS is better suited to the determination of the lanthanide series of elements in many geological applications because **the mass spectrum** is much simpler than the equivalent optical spectrum.

Sample preparation methods are similar to those used for FAAS and ICP-AES. However, **nitric acid** is favoured for sample digestion since the other mineral acids contain elements which cause spectroscopic inteferences.

Because of its capability for rapid multielement analysis, ICP-MS is particularly suited to sample introduction methods which give rise to transient signals. For example, electrothermal vaporization, flow injection and chromatographic methods can be interfaced and many elements monitored in a single run (see Chapter 7).

A major attraction is the ability to perform **isotope ratio** measurements, e.g. in many geological applications to determine the age of rocks, and **isotope dilution analysis**. The latter in particular is gaining-popularity as a highly **accurate, precise** and hence **traceable**, method of analysis, so it is worthwhile describing these techniques in more detail.

## 5.7.1  Isotope ratio analysis

Isotope ratio measurements are performed whenever the exact ratio, or **abundance**, of two or more isotopes of an element must be known. For example, the isotopic ratios of lead are known to vary around the world, so it is possible to determine the source of lead in paint, bullets and petrol by knowing the isotopic abundances of the four lead isotopes 204, 206, 207, 208. Another example is the use of **stable isotopes** as metabolic tracers, where an animal is both fed and injected with an element having artificially enriched isotopes and the fractional absorption of the element can be accurately determined.

In order to perform the isotope ratio experiment correctly it is necessary to compensate for a number of **biases** in the instrumentation. Quadrupole mass spectrometers and their associated ion optics do not transmit ions of different mass equally. In other words, if an elemental solution composed of two isotopes with an exactly 1:1 ratio is analysed using ICP-MS, then a 1:1 isotopic ratio will not necessarily be observed. In practice, transmission through the quadrupole increases up to the mid-mass range (ca $m/z$ 120), then levels off or decreases gradually up to $m/z$ 255. This so-called **mass bias** will differ depending on mass, with the greatest effects occurring at low mass, the least effect in the mid-mass range and intermediate effects at high mass, as shown in Fig. 5.10. Even very small mass biases can have deleterious effects on the accuracy of isotope ratio determinations, so a correction must always be made using an **isotopic standard** of known composition, as shown in the equation

$$C = \frac{R_t}{R_o} \tag{5.3}$$

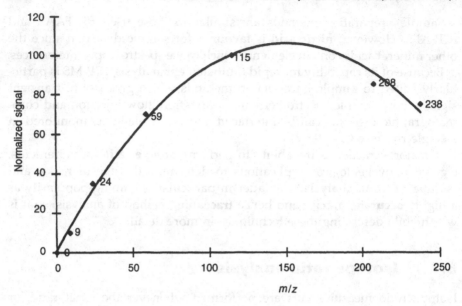

**Figure 5.10** The relative response for equimolar concentration of elements of different mass, corrected for abundance. This reflects the variable transmission of ions through the ion lenses and quadrupole.

where $C$ = mass bias correction factor, $R_t$ = true isotopic ratio for the isotope pair and $R_o$ = observed isotope ratio for the isotope pair.

For the best results, such a correction should be applied to each individual isotopic pair that is to be ratioed, although this is not always possible in practice since a large number of isotopic standards would be needed to cover every eventuality.

Also important is the effect of **detector dead time**. When ions are detected using a pulse counting (PC) detector, the resultant electronic pulses are approximately 10 ns long. During and after each pulse there is a period of time during which the detector is effectively 'dead' (i.e. it cannot detect any ions). The dead time is made up of the time for each pulse and recovery time for the detector and associated electronics. Typical dead times vary between 20 and 100 ns. If dead time is not taken into account there will be an apparent reduction in the number of pulses at high count rates, which would cause an inaccuracy in the measurement of isotope ratios when abundances differ markedly. However, a correction can be applied as follows:

$$C_t = \frac{C_o t}{t - C_o D} \tag{5.4}$$

where $C_t$ = true number of counts, $C_o$ = observed number of counts, $t$ = dwell time (the time spent monitoring each mass) and $D$ = dead time defined in configuration software. The corresponding equation for count rate is

$$R_t = \frac{R_o}{1 - R_o D} \tag{5.5}$$

where $R_t$ = true count rate and $R_o$ = observed count rate.

If no dead time correction is applied, then a linear calibration would not be possible, since the higher count rates between $10^4$ and $10^6$ Hz would be underestimated. This provides a way of determining the dead time empirically, i.e. by re-integrating the isotopic signals with different dead times until a linear calibration is obtained for a series of accurately known standards.

A better method, which accounts for any instrumental drift, is to measure the isotope ratio of two isotopes of an element with different abundances, such as In, Pb or Rb, and use the following expression derived from Eqn. 5.5:

$$(R_m - CR_M) = D[R_m R_M (1 - C)] \tag{5.6}$$

where $R_m$ = count rate of minor isotope, $R_M$ = count rate of major isotope and $C = R_m/R_M$. The isotopic ratio $C$ must be calculated in absence of dead time effects (i.e. at low count rates, but not so low as to give bad counting statistics) by repeated measurements of the blank-subtracted isotope ratio. This is the instrumental isotope ratio, and no correction is made for mass bias. The count rates are then measured for a series of concentrations and the data, which have not been corrected for dead time, can be used to plot $R_m - CR_M$ against $R_m R_M (1 - C)$, the slope of the line being equivalent to the dead time, $D$, in seconds. The data which are plotted must fall within the range of count rates at which dead time effects become significant (i.e. between ca $10^3$ and $10^6$ Hz), otherwise a curve rather than a straight line will result. In practice, the effect of dead time will not be significant as long as the count rate is below $10^5$ counts $s^{-1}$. More important is the effect of low count rate on precision, which means that the longest possible time should be allowed for ion counting when the count rate is below ca 500 count $s^{-1}$, although this will depend on the amount of sample that is available.

The **precision** of any isotope ratio measurement is heavily dependent on the operating conditions of the quadrupole mass filter, in particular the rate at which it hops between masses (**peak hopping**), the time it spends monitoring each mass (the **dwell time**) and the total time spent acquiring data (total **counting time**). A long total counting time is desirable, because the precision is ultimately limited by **counting statistics** and the more ions that can be detected the better. A rapid peak-hopping rate is also desirable in order to eliminate the effects of **drift** (i.e. short-term fluctuations in sample transport, ionization efficiency and ion extraction) inherent in the sample introduction system and the ICP source. The best conditions must be determined empirically for each individual instrument.

## 5.7.2 Isotope dilution analysis

A direct result of the ability to measure isotope ratios with ICP-MS is the technique known as **isotope dilution analysis**. This is done by spiking the sample to be analysed with a known concentration of an enriched isotopic standard, and the isotope ratio is measured by mass spectrometry. The observed isotope ratio ($R_m$) of the two chosen isotopes can then be used in the isotope dilution equation (Eqn. 5.7) to calculate the concentration of the element in the sample:

$$R_m = \frac{A_x C_x W_x + A_s C_s W_s}{B_x C_x W_x + B_s C_s W_s} \tag{5.7}$$

where: $R_m$ = observed isotope ratio of A to B, $A_x$ = atom fraction of isotope A in sample, $B_x$ = atom fraction of isotope B in sample, $A_s$ = atom fraction of isotope A in spike, $B_s$ = atom fraction of isotope B in spike, $W_x$ = weight of sample, $W_s$ = weight of spike, $C_x$ = concentration of element in sample and $C_s$ = concentration of element in spike.

Equation 5.7 can be rearranged into the appropriate form to calculate $C_x$ as follows:

$$C_x = \frac{C_s W_s}{W_x} \times \frac{A_s - R_m B_s}{R_m B_x - A_x} \tag{5.8}$$

The concentration of individual isotopes can then be calculated simply by multiplying the elemental concentration by the abundances of the respective isotopes.

The best precision is obtained for isotope ratios near unity (unless the element to be determined is near the detection limit, when the ratio of spike isotope to natural isotope should be between 3 and 10) so that noise contributes only to the uncertainty of natural isotope measurement. Errors also become large when the isotope ratio in the spiked sample approaches the ratio of the isotopes in the spike (overspiking), or the ratio of the isotopes in the sample (underspiking), the two situations being illustrated in Fig. 5.11. The accuracy and precision of the isotope dilution analysis ultimately depend on the accuracy and precision of the isotope ratio measurement, so all the precautions that apply to isotope ratio analysis also apply in this case.

Isotope dilution analysis is attractive because it can provide very **accurate** and **precise** results. The analyte acts as its own *de facto* internal standard. For instance, if the isotopic spike is added prior to any sample preparation,

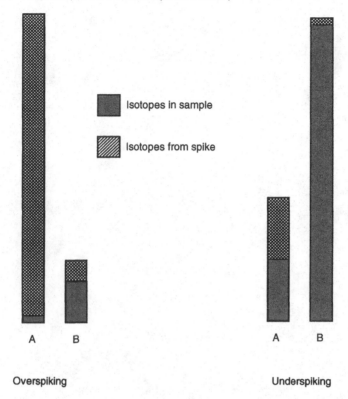

Isotopes in sample

Isotopes from spike

A        B                                              A        B

Overspiking                                            Underspiking

**Figure 5.11**   Representation of overspiking and underspiking in an isotope dilution experiment.

then the spike will behave in exactly the same way as the analyte because it is chemically identical[†]. Hence, any losses from the sample can be accounted for because the analyte and spike will be equally affected.

**Q.**   What are the factors which most affect the accuracy and precision of isotope ratio measurements?

**Q.**   What are the advantages of isotope dilution analyses compared with external calibration?

[†] This is not always true because the analyte could have a different oxidation state to the spike, or be associated with compounds in the sample. Hence it is often necessary to ensure that the spike is equilibrated with the sample by subjecting it to acid digestion or repeated oxidation/reduction.

# 6 ATOMIC FLUORESCENCE SPECTROMETRY

Atomic fluorescence spectrometry is based on the absorption of optical radiation of suitable frequency (wavelength) by gaseous atoms and the resultant **deactivation** of the excited atoms with the release of radiation. The frequencies (wavelengths) emitted are characteristic of the atomic species.

Although AFS combines, for some elements, the advantages of the **large linear dynamic range** typical of atomic emission techniques and the high **selectivity** of atomic absorption, the method has so far not found widespread use.

## 6.1 THEORY

The six types of flame AFS which have been observed are summarized in Fig. 6.1. **Resonance fluorescence** is generally the most useful type because it generates the most intense fluorescence. It occurs when the atom re-emits a spectral line of the same wavelength as that used for excitation. **Direct line fluorescence** has also been exploited analytically. It occurs when transitions between the excited state of the resonance line and a lower intermediate level are allowed by the selection rules. As a result, this type of fluorescence occurs at a higher wavelength than the excitation radiation. The advantage of this technique is that if appropriate filters are used, scatter from the excitation radiation can be eliminated. **Stepwise line fluorescence** occurs when an atom is excited to a higher state by radiation and then undergoes a partial deactivation (by collision or other radiationless process) to a lower state rather than return directly to the ground state. If a transition then occurs from this level to a still lower level (often the ground state), stepwise line fluorescence is observed.

**Thermally assisted fluorescence** is the converse of stepwise line fluorescence. It occurs as a result of a stepwise absorption of energy by

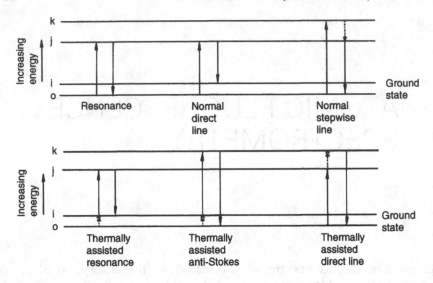

**Figure 6.1** Types of atomic fluorescence. The solid lines represent radiational processes and the dashed lines non-radiational processes. In the latter, a single-headed arrow represents non-radiational deactivation and a double-headed arrow a thermal activation process. The term anti-Stokes is used when the radiation emitted is of shorter wavelength, i.e. greater energy than that absorbed.

an atom to reach an excited state. A simple mechanism is that by which the atom is thermally excited by the atom cell (flame) into a more excited state. Absorption of radiation by this excited state will excite the atom to an even higher state. Radiation emitted from this state is termed stepwise excitation fluorescence. It occurs most frequently when the first excited state is very close to the ground state, although it can occur elsewhere if an extremely hot atom cell is used. This type of fluorescence is weak though and has found few analytical applications. It can give rise to **anti-Stokes fluorescence**, i.e. when the light emitted from the atom is at a shorter wavelength than that used to excite it.

The intensity of the atomic fluorescence radiation, $I_F$, is proportional to the intensity of the absorbed radiation, which when using a line source is directly proportional to the source intensity, $I_S$.

At low concentrations, $I_F$ is proportional to the concentration of the analyte; at higher concentrations, **self-absorption** is observed (cf. AES). This behaviour is reflected in the growth curves of $\log I_F$ versus $\log N$. Figure 6.2 is based on these results. These curves often show linearity of four orders of magnitude before curvature.

As $I_F$ depends on $I_S$, a stable, **intense sharp-line source** greatly enhances AFS sensitivity. Similarly, the geometry of the atom cell is important.

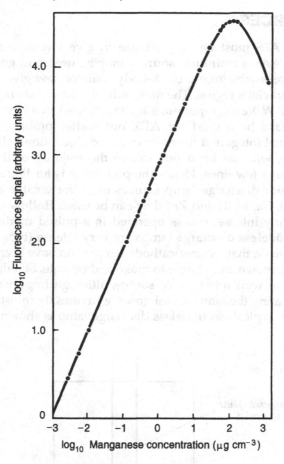

**Figure 6.2** Growth curve for manganese atomic fluorescence at 280 nm.

## 6.2 INSTRUMENT DESIGN

In some flame AFS systems, **interference filters** and **solar blind photomulti-pliers** have been used to reduce the background, but usually a conventional monochromator is used. As in AAS, the source signal is **modulated** so that the atomic fluorescence can be distinguished from atomic emission.

It has been shown that a high frequency of modulation of the **electrodeless discharge lamp** (e.g. 10 kHz) is advantageous. This frequency is well away from the low frequency of flame noise. If the amplifier is 'locked-in' to this high frequency via a reference signal, an optimum signal-to-noise ratio is achieved.

## 6.3  SOURCES

A source for AFS must be very intense to give low detection limits. In contrast with AAS, **continuum sources** may be used but, given the distribution of wavelengths from a black-body radiator, few give high intensity in the vital ultraviolet region. The most suitable continuum radiation source is the 150–500 W **Xe high-pressure arc** lamp, used in a continuous mode. Xenon arcs have been used for AFS, but **scatter** problems are encountered. The actual integrated intensity over the absorption half-width is still relatively low, whereas for a line source the majority of the emission is concentrated in a few lines. Hence the preference is for line sources.

Where **vapour discharge lamp** sources exist (for volatile elements such as Hg, Na, Cd, Ga, In, Tl and Zn) they can be used. **Hollow-cathode** lamps are insufficiently intense, unless operated in a pulsed mode. **Microwave-excited electrodeless discharge** lamps are very intense (typically 200–2000 times more intense than hollow-cathode lamps) and have been widely used. They are inexpensive and simple to make and operate. Stability has always been a problem with this type of source, although improvements can be made by operating the lamps in microwave cavities thermostated by warm air currents. A typical electrodeless discharge lamp is shown in Fig. 6.3.

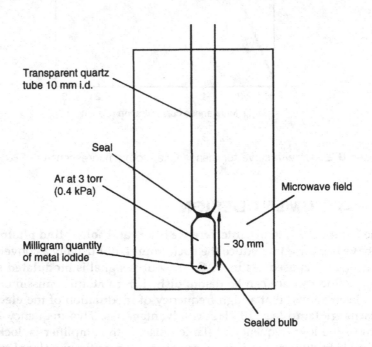

**Figure 6.3**  A microwave-excited electrodeless discharge lamp.

The high intensity of **lasers** makes them obvious, but expensive, candidates as sources for AFS, and instruments for laser-induced atomic fluorescence spectrometry (LIF) have been developed in recent years. The advent of the tuneable dye laser led to the possibility of selecting wavelengths from a laser, and frequency doubling (second harmonic generation) has permitted the excitation of AFS lines in the ultraviolet region (Ca 220 nm using a $N_2$ laser pump or Ca 180 nm with an Nd:YAG laser pump). The light from such a source can be sufficiently intense to cause **saturation fluorescence** (i.e. population inversion), thus nullifying the effects of quenching (e.g. radiationless return to the ground state induced by flame species at flame temperatures) and self-absorption.

## 6.4  ATOMIZATION

### 6.4.1  Flames

The fluorescence power yield is always less than unity. This **non-radiative loss** of energy is referred to as 'quenching'. Quenching increases with temperature (the number of collisions) and quenching cross-section of the colliding particle (argon has a negligible, hydrogen a low, oxygen a high quenching cross-section). The ideal atom cell for AFS would also exhibit no background, thus allowing the detection of very small signals.

There has been interest in the low radiative background, low quenching argon–hydrogen diffusion flame. The temperature of this flame is too low to prevent severe **chemical interferences** and therefore the argon-separated air–acetylene flame has been most widely used. The hot nitrous oxide–acetylene flame (argon separated) has been used where atomization requirements make it essential. In all cases, circular flames, sometimes with mirrors around them, offer the preferred geometry.

Elements such as As, Se and Te can be determined by AFS with hydride sample introduction into a flame or heated cell followed by atomization of the hydride. Mercury has been determined by cold-vapour AFS. A **non-dispersive** system for the determination of Hg in liquid and gas samples using AFS has been developed commercially (Fig. 6.4). Mercury ions in an aqueous solution are reduced to mercury using tin(II) chloride solution. The mercury vapour is continuously swept out of the solution by a carrier gas and fed to the fluorescence detector, where the fluorescence radiation is measured at 253.7 nm after excitation of the mercury vapour with a high-intensity mercury lamp (detection limit 0.9 ng $l^{-1}$). Gaseous mercury in gas samples (e.g. air) can be measured directly or after **preconcentration** on an absorber consisting of, for example, gold-coated sand. By heating the absorber, mercury is **desorbed** and transferred to the fluorescence detector.

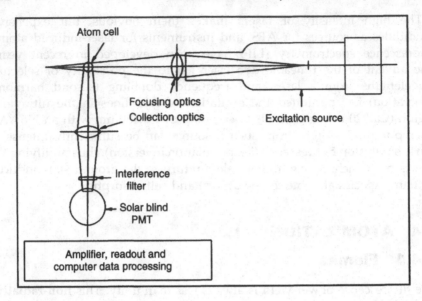

Atom cell

Focusing optics
Collection optics

Excitation source

Interference
filter

Solar blind
PMT

Amplifier, readout and
computer data processing

**Figure 6.4** Non-dispersive AFS system (with permission from PS Analytical Ltd).

This preconcentration method can also be applied to improve the detection limits when aqueous solutions (or dissolved solid samples) are analysed.

## 6.4.2 Electrothermal atomizer

Electrothermal atomizers are also suitable for AFS as, when an inert gas atmosphere is used, quenching will be minimized. In the nuclear, electronic, semiconductor and biomedical industries where detection limits have to be pushed as low as 1 part in $10^{13}$ (or 0.1 pg g$^{-1}$ in the original sample), electrothermal atomization with a laser as excitation source (LIF-ETA) may be used. Figure 6.5 shows schematically a common way of observing the fluorescence in LIF-ETA. The fluorescence signal can be efficiently collected by the combination of a plane mirror, with a hole at its centre to allow excitation by the laser, positioned at 45° with respect to the longitudinal axis of the tube and a lens chosen to focus the central part of the tube into the entrance slit of the fluorescence monochromator.

**Q.** Which type of AFS is of greatest practical use?

**Q.** Can sensitivity be improved by increasing the intensity of the source in (i) AAS and (ii) AFS?

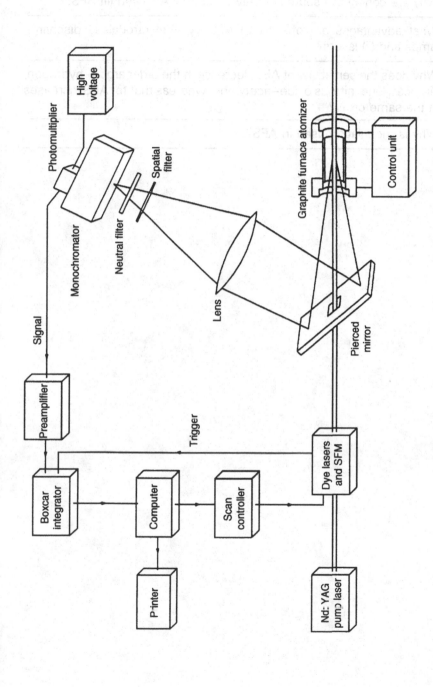

**Figure 6.5** Laser-induced fluorescence with electrothermal atomization (reproduced with permission from Heitman *et al.*, *J. Anal. At. Spectrom.*, 1993, **9**, 437).

**Q.**  Why are continuum sources rarely used in (i) AAS and (ii) AFS?

**Q.**  What advantages are offered to AFS by (i) electrodeless discharge lamps and (ii) lasers?

**Q.**  Why does the sensitivity of AFS decrease in the order argon–hydrogen, air–acetylene, nitrous oxide–acetylene, whereas that for AES increases in the same order?

**Q.**  Why is modulation used in AFS?

# 7 SPECIAL SAMPLE INTRODUCTION TECHNIQUES

In addition to conventional aspiration, using a nebulizer and spray chamber, samples may be introduced in to atomic spectrometers in a number of different ways. This may be because a knowledge of **speciation** (i.e. the organometallic form or oxidation state of an element) is required, to introduce the sample while minimizing interferences, to increase sample transport efficiency to the atom cell or when there is a limited amount of sample available.

## 7.1 PULSE NEBULIZATION

For samples that are limited in volume, that contain high dissolved solids or large amounts of organic solvents that may theoretically cause spectroscopic interferences, **pulse nebulization** (a very basic form of flow injection) may be used. This procedure permits the use of smaller samples (25–200 mm$^3$) and higher concentrations (e.g. 10% w/v steel) than in normal nebulization. A cup, made of **Teflon** or other suitable plastic, is attached to the nebulizer tube, and the sample **pipetted** into the cup using a micro-pipette. The sample is totally consumed and a peak signal is observed. For example, the technique may be used for steel analysis or in a biological study where only a small amount of sample is available. The technique also enjoys a number of alternative names: discrete sample nebulization, gulp sampling, direct-injection cup nebulization, and Hoechst cup nebulization. The technique has been used mainly for sample introduction to flame spectrometry.

## 7.2 OTHER DISCRETE SAMPLING DEVICES

There have been two other discrete sampling devices that have been used in conjunction with flame spectroscopy. The **Kahn sampling boat** is a system whereby nebulizer inefficiency was avoided by using a **tantalum sampling boat** from which the sample was evaporated as it was pushed into the flame. The technique was only applicable to the more **easily atomized** analytes, but it did yield a useful improvement in sensitivity. Unfortunately, the technique is prone to poor reproducibility. The **Delves sampling cup** was a modification of the above technique. It was applied mainly to the determination of lead in **microsamples** of whole blood or other easily combustible material. The tantalum cup was replaced with a smaller (10 mm outer diameter, 5 mm deep, 0.15 mm thick metal foil) and more easily positioned nickel or stainless-steel micro-crucible. The crucible may also be used during preliminary chemical treatment (e.g. addition of hydrogen peroxide to destroy organic matter). It is mounted into a device which enables it to be pushed close to the flame to allow charring, and then into the flame, to allow atomization. A nickel (or, more recently, silica) **absorption tube** is mounted in the flame and the atoms enter the tube through a hole half way along its length. Light passes through the tube, thus improving the reproducibility by lengthening and defining the residence of the atoms in the flame. Both of the techniques described above have fallen into disuse, mainly because of the advent of flow injection.

## 7.3 SLOTTED TUBE ATOMIZER

Although the **slotted tube atomizer** is not strictly a sample introduction technique, it should be mentioned as it has found use in a number of different sample introduction techniques for flame spectroscopy. The majority of techniques with which it has been used are those that produce a **transient signal** or where a long **residence time** for the atoms in the light beam is required owing to the very low concentration of the analyte in the sample. Normally, the atoms pass through the flame extremely quickly and hence spend a very short period of time in the light beam ($10^{-4}$ s), but the tube atomizer retains the atoms for a far longer period and hence the sensitivity increases.

## 7.4 FLOW INJECTION

Flow injection has been coupled with flame-and plasma-based instruments. It has a number of advantages over conventional nebulization, including the ease of coupling, the possibility of **preconcentration** and **matrix removal**

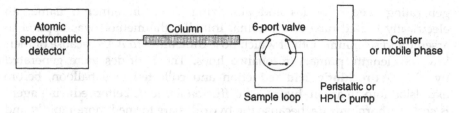

**Figure 7.1** Flow injection manifold for coupling with atomic spectrometry. The column can be used for preconcentration, matrix removal or chromatography.

and the advantages described for pulse nebulization. The introduction of organic solvents into plasma instruments can lead to high reflected power (i.e. power that is reflected back to the generator and which can cause damage), plasma cooling, interference effects and possible plasma extinction. The use of flow injection enables only very **small volumes** to be introduced and hence the deleterious effects are minimized. For some samples, e.g. sea-water, the matrix constituents are very high (e.g. 3.5% w/v salt) and many of the analytes of interest are at very low concentration (ng ml$^{-1}$ level or below). By using a **micro-column** containing an appropriate ion or **chelating exchange resin**, the analytes may be retained while the bulk matrix is allowed to pass to waste. The analytes may then be eluted to detection by the introduction of a small volume of **eluent**. If 20 ml of sample are passed through the column and, after washing with de-ionized water to remove the remnants of the matrix, the analytes are eluted to detection by 0.5 ml of eluent, a preconcentration factor of 40 is achieved. The use of such mini-columns of exchange media is becoming more popular as it facilitates the analysis by both preconcentrating the sample while simultaneously removing potential interferences. Such a technique enables ultra-trace analytes to be determined using simpler and cheaper instrumentation, e.g. a flame spectrometer. If it is coupled with ICP-MS, detection limits substantially below the ng ml$^{-1}$ level may be obtained, even for very complex samples. A typical flow injection manifold that would permit preconcentration or matrix removal is depicted in Fig. 7.1.

## 7.5 VAPOUR GENERATION

### 7.5.1 Hydride generation

The elements Ge, Sn, Pb, As, Sb, Bi, Te and Se, form **covalent gaseous hydrides**. A number of these elements are now routinely determined by generating their hydrides and atomizing these, in either a flame, an electrically heated tube or a plasma. Initially, the method was applied to arsenic and selenium, both of which pose difficulties in AAS because of their low-wavelength, primary resonance lines. The hydrides were generated by **zinc–hydrochloric acid** reduction and collected in a balloon, before expulsion into an argon–hydrogen diffusion flame. A better reducing agent is **sodium borohydride**, because the hydrides are formed more rapidly and a collection reservoir is not needed. A 1% w/v aqueous solution is usually sufficient, provided that there is vigorous stirring of the acidified sample. This reagent can be used to generate the hydrides of antimony, arsenic, bismuth, germanium, lead, selenium, tellurium and tin. The replacement of inefficient nebulization by gaseous sample transport improves the detection limit of all the elements except lead (the improvement for germanium is not dramatic), where the limiting factor is presumably the difficulty in forming plumbane. If the lead is oxidized prior to reduction, improvements in detectability are also observed with this element.

Arsenic and antimony also give responses which vary according to **valence state**, the +5 state giving poorer responses than the +3 state. These elements are therefore **reduced** in iodide solution or by L-cysteine prior to hydride generation. Selenium and tellurium in the +6 state must be reduced to the +4 state prior to hydride generation because the +6 state does not form a hydride. This is normally achieved by heating with hydrochloric acid. This process is often accelerated by the use of microwave technology. It must also be noted that many of the arsenic and selenium compounds that occur naturally in nature do not form hydrides. Examples include arsenobetaine (a very common arsenic-containing compound in marine organisms) and selenomethionine. These compounds are very stable, even to attack by concentrated nitric acid, and must be broken down to arsenic and selenium ions by much harsher conditions, e.g. perchloric acid or by photolysis. Severe underestimates of arsenic and selenium contents are often obtained because the analyst has failed to destroy completely non-hydride forming compounds such as these.

Considerable inter-element interference effects have been reported on the actual hydride-forming step. Elements easily reduced by sodium borohydride (e.g. silver, gold, copper, nickel) give rise to the greatest **suppressions**. These interfering ions may be removed by the addition of **masking agents** that complex with them.

## 7.5.1.1 Instrumentation

Hydride generation (HG) systems make use of a **gas–liquid separator** to separate the gaseous hydrides from the liquid reagents prior to introduction into the atom cell. The process can be performed in **batch mode** or in **continuous operation**.

In most arrangements used for AAS, the batch mode of operation is preferred (Fig. 7.2a). The acidified sample is transported into a stirred glass cell containing 1% w/v aqueous sodium borohydride. The contents are mixed and the liberated hydrides flushed with an inert gas into either

(i)   a flame, often an argon–hydrogen diffusion flame;
(ii)  a narrow-bore silica tube mounted over an air–acetylene flame (in one design there is a transverse flow of nitrogen at the ends of the tube to ensure the liberated hydrogen does not burn in the light path — some results from such a system are shown in Table 7.1);
(iii) a narrow-bore silica tube, electrically heated by means of a winding of suitable resistance wire.

The use of narrow-bore tubing results in much improved limits of detection by limiting the dilution of the hydrides. Using arrangement (ii) or (iii), background correction is usually unnecessary, provided that hydrogen is not allowed to burn in the optical axis.

Hydrides may also be determined using **atomic fluorescence** detectors. Several commercial instruments are now available that specialize in the determination of specific analytes. One example is an HG-AFS system for the determination of As and Se.

The introduction of hydrides into plasma-based instrumentation has also been achieved. The sensitivity increases markedly when compared with conventional nebulization because of the **improved transport efficiency** of the analyte to the atom cell (close to 100%). Often, a membrane gas–liquid separator is used to ensure that aerosol droplets of liquid do not reach the plasma.

**Table 7.1** Typical figures of merit for HG-FAAS.

| Element | Wavelength (nm) | Reductant | Characteristic concentration ($\mu$g cm$^{-3}$) | Detection limit ($\mu$g cm$^{-3}$) |
|---|---|---|---|---|
| As | 193.7 | NaBH$_4$ | 0.00052 | 0.0008 |
|    |       | Zn-HCl  | 0.001   | 0.0015 |
| Bi | 223.1 | NaBH$_4$ | 0.00043 | 0.0002 |
| Ge | 265.1 | NaBH$_4$ | 1.0     | 0.5    |
| Pb | 283.3 | NaBH$_4$ | 0.08    | 0.1    |
| Sb | 217.6 | NaBH$_4$ | 0.00061 | 0.0005 |
| Se | 196.1 | NaBH$_4$ | 0.0021  | 0.0018 |
| Sn | 224.6 | NaBH$_4$ | 0.00044 | 0.0005 |
| Te | 214.3 | NaBH$_4$ | 0.002   | 0.0015 |

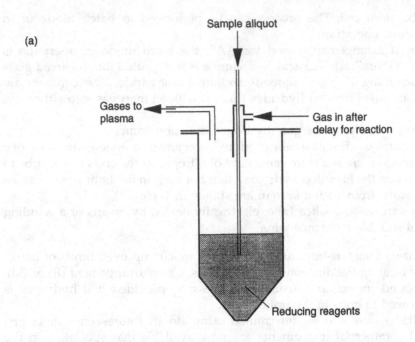

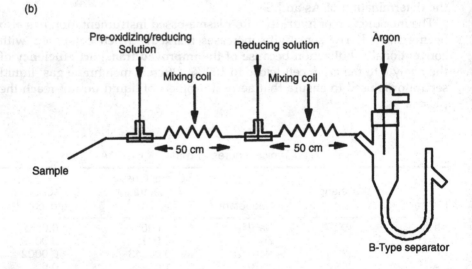

**Figure 7.2** Hydride generation systems: (a) batch mode operation; (b) continuous mode operation.

## 7.5.2  Cold vapour generation

This applies solely to **mercury** as it is the only analyte that has an appreciable atomic vapour pressure at room temperature. The 253.7 nm line is usually used for mercury atomic absorption, but the transition is spin forbidden, and relatively insensitive. The 184.9 nm line is potentially 20–40 times more sensitive, but at this wavelength most flame gases and the atmosphere absorb strongly. Thus, flame methods for mercury are not noted for their sensitivity (typical flame detection limits are in the range 1–0.1 µg ml$^{-1}$). If chemical reduction is employed, mercury can be brought into the vapour phase without the need to use a flame, and detection limits are dramatically improved.

Four main methods have been used to bring mercury into the vapour phase:

(i)   Reduction–aeration: this is by far the most common of the methods used to bring mercury into the vapour phase. Mercury in aqueous solution is treated with a **reducing agent** and then swept out of solution by bubbling a gas through it. The most typical reducing agent used is **tin(II) chloride**, although tin(II) sulphate has also been used. A few workers have used sodium borohydride, but this can pose safety problems, and the evolution of large amounts of hydrogen increases the quenching if AFS detection is employed.

(ii)  Heating: the sample is **pyrolysed** and combusted.

(iii) Electrolytic amalgamation: mercury is plated on to a copper cathode during **electrolysis**. The cathode is then heated as in (ii), to release the mercury.

(iv)  Direct amalgamation: mercury is collected on a **silver** or **gold wire**, from which it is released by heating. This method may be employed with (i) or (ii) as a concentration method. Specialist commercial instruments have been developed that pyrolyse samples and then **trap** the mercury evolved on gold.

A system for cold vapour AAS is shown in Fig. 7.3. The evolved mercury vapour is passed to a long path-length **absorption cell**, usually constructed of Pyrex glass tubing with silica end windows. A transient absorption peak is observed. In some systems, a **recirculating pump** is used to cycle the mercury vapour around the system and achieve a steady reading.

Such arrangements as in Fig. 7.3 are exceedingly simple and can be built 'in-house'. A hollow-cathode lamp is used as a source and the cell can be taped to the end of the burner. Air from a small compressor can be used for aeration. As an enclosed cell is used, problems can arise from water condensation and spray. The cell can be heated to minimize such

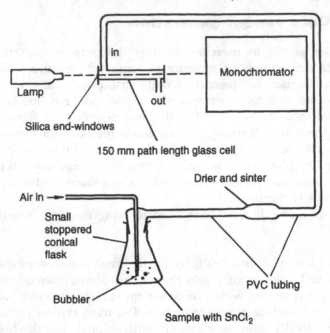

**Figure 7.3** Typical cold vapour generation AAS system used for mercury determination. The same system can be used with a flame in place of the Pyrex tube to allow the determination of hydride-forming elements.

problems. Spurious absorption signals can often be seen and the use of background correction is to be preferred. Detection limits of about 1 ng are achievable. Certain elements, particularly silver, gold, palladium and platinum, are reported to **cause chemical interferences** in the reduction stage. Specialist fluorescence-based instrumentation has also been developed for the determination of mercury. Argon is used as both **a sheathing gas** to enclose the plume of mercury and as the flushing gas to transport it to the detector. This is because argon causes less **quenching** of the flame than nitrogen.

## 7.5.3 Other vapour techniques

There have been a few research papers reporting the use of other vapour generation techniques to volatilize analytes that form unstable hydrides, or had previously been thought not to form vapours at room temperature. Examples include the use of **sodium tetraethylborate** to form volatile ethyl compounds of cadmium, lead and thallium.

## 7.6  CHROMATOGRAPHY

Several different types of **chromatography** have been coupled with atomic spectrometric detectors. Most applications involving chromatography coupled with atomic spectrometry yield **speciation** data, i.e. they separate different chemical forms of an analyte.

Liquid chromatography is the most common type because it is so easily coupled with the majority of atomic spectrometric detectors. The **flow rates** associated with ion chromatography or HPLC (typically 1–2 ml min$^{-1}$) are compatible with the sample uptake rate of inductively coupled plasma-based instruments. However, the flow rate is too low to be coupled directly to a flame spectrometer, whose uptake is typically 6–10 ml min$^{-1}$. For this coupling to be made, it is frequently necessary to have a connection that contains a small opening to the atmosphere so that air may be entrained, thus preventing the formation of a vacuum. Other components of the **coupling** may be required depending on the nature of the chromatography. If organic solvents are used in the mobile phase (e.g. methanol, acetonitrile) then these have the effect of quenching plasmas and, under extreme conditions, may lead to plasma extinction. A **desolvation** device, e.g. a **membrane drier** tube or **Peltier** cooler, is therefore necessary. Coupling of liquid chromatography with microwave-induced plasmas is still in the research stages, because the plasma is so easily extinguished.

Gas chromatography has been coupled to atomic spectrometric detectors, but the coupling is less straightforward than for liquid chromatography. This is because the analytes must be transferred from the GC oven to the atom cell at an **elevated temperature** to prevent analyte condensation (loss of analyte and therefore loss of sensitivity). A number of research papers have been published that describe the construction of such transfer lines. For flame detection, a slotted tube atomizer or some other quartz or ceramic tube is usually placed in the flame to improve sensitivity. For inductively coupled plasma work, the transfer line usually leads directly to the back of the torch. Care must be taken to ensure that the transfer line does not act as an aerial for the RF power. This could result in damage to instrumentation and to operators. The positioning of the transfer line within the torch and the sheathing gas flow should be optimized very carefully, since these parameters have a dramatic effect on the sensitivity. A commercial piece of instrumentation is available that has coupled GC with MIP-AES detection (see Chapter 4). As described previously, the MIP is very easily extinguished, and even microlitre quantities of organic solvent frequently have to be vented from the system before the analyte reaches the plasma. Other types of chromatography, e.g. **capillary electrophoresis**, are yet to make a full impact on coupled techniques, but research is on-going.

## 7.7 SOLID SAMPLING TECHNIQUES

There are a variety of solid sampling techniques that have been used to introduce samples. **Electrothermal vaporization** of a solid or liquid sample from a rod or a tube into plasma instruments yields detection limits far superior to those obtained by conventional nebulization. Sample transport approaches 100% and, if a solid sample is vaporized, the dilution factors associated with the dissolution of the sample are not present. In addition, interferences arising from constituents of the matrix, e.g. chlorides, oxides, may be removed during the **charring stage** (as in ETAAS), thereby facilitating interference-free determination of the analyte when it is vaporized into the plasma. The process is very similar to that for ETAAS. Sample is dispensed on to the vaporisation device and it then undergoes a temperature programme so that it is dried, ashed and then vaporized. A sheath of gas (typically argon) then transports the analyte to the plasma, where it is detected as a transient peak. For some analytes (notably those that form very refractory oxides), the addition of a **matrix modifier** assists in the vaporization of the analyte into the plasma. Fluoride-based modifiers such as **PTFE** (in the form of a suspension) or **Freon gases** have proved very popular, because they vaporise these analytes as their **fluorides** at a much lower temperature than in the modifier's absence. This helps to prolong the lifetime of the electrothermal device in addition to improving the transport of the analyte to the plasma. The plasma is energetic enough to dissociate the analyte fluorides, so that analyte atoms (or ions) may be detected.

Another popular method of solid sample introduction is **laser ablation**. Here, a laser (often an Nd:YAG laser) is focused on to the surface of a sample (although occasionally focusing either just above or below the surface is used). **Pulsing** of the laser leads to localised vaporization of the sample, which can then be transported to the atom cell by a flow of inert gas. This technique avoids the need for sample dissolution and so yields good detection limits. Laser ablation has several advantages over other introduction techniques. It is particularly useful when only a very limited amount of sample is available. The laser can be focused on to a very small area so that **single crystals/particulates** may be analysed. By repetitive ablation of the same spot, **depth profiling** of a sample may also be made. The main drawback with the technique is that it can be very difficult to **calibrate** accurately. This problem arises from the fact that very closely **matrix-matched** standards must be used to calibrate, and it can be hard to find samples of known composition that are matched closely enough. Despite this problem, many workers have succeeded in analysing a variety of matrices. Another problem arises from the small sampling area (ca 50 μm diameter spot), so that a representative sample is not obtained unless the sample is very

**homogeneous.** However, this can be an advantage if a **spatial map** of the analyte within the sample is required.

**Slurry nebulization** has also proved very popular. In this technique, sample (typically 0.25 g) is placed in a 30 ml plastic bottle and 10 g of expanded zirconia beads are added. A **dispersant** is added and the bottle is sealed and then placed on a mechanical shaker for several hours. During the shaking, the zirconia beads **grind** the sample into **very fine particles**. After dilution to a known volume, the slurry may be aspirated directly into an atomic spectrometric instrument. Other methods of slurry preparation also exist, e.g. using a micronizer, but the 'bottle and bead' method is the most common.

There are several advantages to slurry nebulization:

(i)   samples may be analysed against **aqueous calibrants**;
(ii)  it avoids the use of hazardous acids that are required for digestion/dissolution;
(iii) it is relatively **contamination free**;
(iv)  it avoids possible losses of volatile analytes.

The main disadvantage of slurry nebulization is that a very small particle size is required. If the particles are too large, then once diluted to volume, the particles sink to the bottom of the flask. The slurry aspirated may therefore not be representative of the original sample. Another problem that arises when the particle size exceeds approximately 2 μm, is that the slurry no longer behaves in a similar way to aqueous solution, and can therefore not be determined against aqueous standards. In the spray chamber of the spectrometer, larger particles pass directly to waste, and therefore a non-representative sample is presented to the plasma. For sample preparation, several dispersants may be used. For inorganic samples, e.g. glasses, firebricks, dispersants such **as sodium hexametaphosphate** or **sodium pyrophosphate** may be used. For samples that are more organic or biological in nature, **Triton X-100** or **Aerosol-OT** are available. Slurries have also been analysed using ETAAS. The preparation is similar, but the particle size is less critical. The slurries have to be stabilized for analysis using an autosampler, i.e. the slurry has to be homogeneous when it is analysed. To ensure this, an **ultrasonic probe** has been developed that mixes the slurry just prior to injection.

# 7.8 NEBULIZERS

In addition to concentric pneumatic nebulizers that are used most frequently for the majority of atomic spectrometric systems, a range of other nebulizers also exist. Some nebulizers that are used for plasma instrumentation, e.g. **direct injection nebulizers** (DIN) or **ultrasonic nebulizers** (USN), increase the sample transport efficiency to the plasma and hence increase the sensitivity. The DIN nebulizer transports close to 100% of the sample to the plasma but it can cope with only very low flow rates of sample uptake (typically $< 100 \, \mu l \, min^{-1}$). It also suffers the disadvantage of becoming **blocked** very easily by particulate matter, or by samples with a high dissolved solid content. Ultrasonic nebulizers are very expensive and also suffer the drawback of having long '**memory effects**'. Various other nebulizer types also exist, including those that specilize in aspirating samples containing high salt content, e.g. **Babington-type**, and the far more exotic types including **thermospray** and **electrospray** sample introduction systems. The nebulizer chosen for an analysis will be governed by the nature of the sample.

# APPENDICES

# APPENDIX A
# REVISION QUESTIONS

1 What is meant by the following terms:
   (a) ground state;
   (b) resonance line;
   (c) excited state;
   (d) absorption line half-width;
   (e) self-absorption;
   (f) Maxwell–Boltzmann distribution;
   (g) atomic fluorescence?

2 Describe the factors which cause broadening of spectral lines. In atomic absorption spectrometry, why is it preferable for the source line-width to be narrower than the absorption profile?
   How can this be achieved?
   What are the differing requirements for resolution in monochromators for atomic emission and for atomic absorption spectrometry?

3 Explain the operation of:
   (a) a photomultiplier tube;
   (b) a pneumatic nebulizer;
   (c) an electrothermal atomizer;
   (d) a hollow-cathode lamp.

4 Describe the principles of:
   (a) atomic emission spectrometry;
   (b) atomic mass spectrometry.
   Discuss the instrumental requirements for these two techniques.

**5** (a) Discuss the importance of the temperature of the atom cell in atomic emission spectrometry and in atomic absorption spectrometry.

  (b) Describe how the requirements of the monochromator and detection system differ in atomic emission and atomic absorption spectrometry.

**6** With regard to both theoretical principles and practical considerations of inductively coupled plasma atomic emission spectrometry, discuss the design of two of the following:

  (a) nebulizer systems;
  (b) the plasma;
  (c) monochromators and detector systems.

**7** Describe suitable instrumentation for sensitive analytical measurements in atomic absorption spectrometry. Include a discussion of the ways in which the atomic population in the atom cell may be maximized and why the light source is always a line source.

**8** In analytical atomic spectroscopy, how are atomic populations usually formed from solutions? In your answer, include an outline of the conventional apparatus and basic processes involved, and explain how the atomic or ionic population may be maximized.

**9** Discuss why the following are often preferred for practical atomic absorption spectrometry:

  (a) a modulated narrow line source;
  (b) a Czerny–Turner monochromator;
  (c) a small graphite tube atomizer;
  (d) microprocessor controlled curvature correction.

**10** Discuss the reasons for the following:

  (a) Compared with flame atomizers, electrothermal atomizers generally result in enhanced sensitivity for atomic absorption spectrometry.
  (b) Increasing the intensity of the source increases the sensitivity in atomic fluorescence spectrometry, but has relatively little effect in atomic absorption spectrometry.
  (c) Compared with flame atomizers, plasma atomizers generally result in enhanced sensitivity for atomic emission spectrometry.
  (d) Metals dissolved in organic solvents can show either enhanced or reduced sensitivity compared with those in aqueous solution.

**11** Describe a typical electrothermal atomizer for atomic absorption spectrometry. Critically compare graphite furnaces, air–acetylene flames, and nitrous oxide flames as atom cells for atomic absorption spectrometry.

**12** 'The ICP provides the most useful atom cell for atomic emission spectrometry'. Critically discuss this statement with particular reference to the analysis of 'real' samples.

**13** Discuss the relative advantages and disadvantages of:
   (a)   atomic emission and atomic mass spectrometry;
   (b)   flames and electrothermal atomizers;
   (c)   hydride generation.

**14** 'There have been claims that of ICP-AES and ICP-MS, one is superior to the other as an analytical technique'. Discuss this statement critically.

**15** Discuss in detail the origins and effects of interferences in ICP atomic emission and ICP mass spectrometry, and describe how they may be minimized or eliminated in practice. Explain why some of these interferences are common to both methods. Illustrate your answer with suitable examples where appropriate.

**16** Discuss, with appropriate reference to basic theory, the reasons for five of the observations in ICP mass spectrometry listed below:
   (a)   Copper can be determined more sensitively than chloride.
   (b)   In the determination of arsenic in an effluent containing large amounts of sodium chloride, the apparent arsenic signal at $m/z$ 75 was reduced when the sample was pretreated by passing it through an anion-exchange column.
   (c)   The lead 204/208 isotope ratio observed for an isotopic standard was 10% less than the certified value.
   (d)   The determination of antimony in a lead sample gave a low result when compared with aqueous standards, but improved when the sample was diluted.
   (e)   Far fewer spectroscopic interferences are observed with a magnetic sector compared to a quadrupole mass analyser.
   (f)   The determination of iron in water is greatly improved by the use of 'cool plasma' conditions.

**17** Outline methods for performing the following determinations (approximate levels of the analytes are given in parentheses):
   (a)   sodium in soil extracts (50 mg $l^{-1}$);
   (b)   manganese in cast iron (0.1%);
   (c)   tetraethyllead in petrol (300 mg $l^{-1}$);
   (d)   arsenic in trade effluent (0.1 mg $l^{-1}$);
   (e)   magnesium in tap water (20 mg $l^{-1}$);
   (f)   cadmium in tap water (1 µg $l^{-1}$);
   (g)   lead in blood (30 µg $l^{-1}$);
   (h)   bismuth in nickel alloys (1 µg $g^{-1}$);
   (i)   mercury in seaweed (50 ng $g^{-1}$).

# APPENDIX B

# PRACTICAL EXERCISES

## B.1 CALCULATIONS

A variety of units are used in practical atomic spectrometry. Instrumental readings for AAS are generally recorded in units of absorbance, and for AES and MS in counts per second (cps). Concentrations are normally expressed as a weight or mass per volume, e.g. $\mu g \ ml^{-1} = \mu g \ cm^{-3} = mg \ l^{-1} = mg \ dm^{-3}$. Very often 1 $\mu g \ cm^{-3}$ is referred to as 1 ppm and so on. The use of the term parts per million (ppm) should be discouraged as it lacks rigour and is ambiguous. For example, if the cadmium content of a 1% solution of steel in aqua regia is reported as 1 ppm, does this mean 1 $\mu g \ cm^{-3}$ in the solution (i.e. 100 $\mu g \ g^{-1}$ in the solid) or 1 $\mu g \ g^{-1}$ in the solid? The term parts per billion (ppb) can cause further confusion because different definitions of billion are used. Generally, the American billion, $10^9$, is intended, i.e. 1 ppb = 1 $ng \ cm^{-3}$ = 1 $\mu g \ l^{-1}$.

**1** The following blank-corrected readings were obtained for the determination of nickel in steel, using nickel standards dissolved in iron solution (10 $g \ l^{-1}$). The determination was performed by atomic absorption spectrometry using an air–acetylene flame and the 232 nm nickel line.

| Nickel concentration (mg $l^{-1}$) | 1 | 2 | 4 | 6 | 8 | 10 | 12 |
|---|---|---|---|---|---|---|---|
| Absorbance | | 0.06 | 0.11 | 0.22 | 0.34 | 0.44 | 0.55 | 0.60 |

Plot the calibration curve for these data and comment upon any observed deviations from linearity.

Find the characteristic concentration (i.e. the concentration corresponding to an absorbance of 0.0044) for nickel (in iron solution) using this instrumentation by (a) interpolation of the graph; (b) calculation from the slope.

If a 1% solution of steel gave an absorbance of 0.36, what would be the concentration of nickel in the solution, and hence the steel sample as a w/w percentage?

(Answers: 0.08 mg $l^{-1}$, 6.5 mg $l^{-1}$, 0.065%)

2 The procedure for the method of standard additions was described in Section 1.4.1. The following data were obtained for the determination of copper in a contaminated stream by inductively coupled plasma atomic emission using the 324.754 nm copper line.

| Stream water + (cm$^3$) | Analysis solution Distilled water (cm$^3$) | + 1 µg cm$^{-3}$ Cu standard (cm$^3$) | Net signal (cps) |
|---|---|---|---|
| 5 | 5 | 0 | 440 |
| 5 | 4 | 1 | 1540 |
| 5 | 3 | 2 | 2640 |
| 5 | 1 | 4 | 4800 |

Plot the emission signal ($y$-axis) against the added copper content of the solution (in µg; $x$-axis), extrapolate the graph and, from the negative $x$-intercept, calculate the concentration of copper in the original stream water sample.

(Answer: 0.08 µg cm$^{-3}$)

3 The calculation of the standard deviation of a series of readings can be used both to give an estimate of the precision and to calculate the limit of detection (see Section 1.4.2). The following data were obtained for cadmium at $m/z$ 111 when spraying a standard solution of 0.1 ng ml$^{-1}$ cadmium, using an inductively coupled plasma mass spectrometer when using a 60 s scan over the range $m/z$ 5–255.

| Sample No. | 1 | 2 | 3 | 4 | 5 | 6 | 7 | 8 | 9 | 10 | 11 |
|---|---|---|---|---|---|---|---|---|---|---|---|
| Signal (CPS) | 510 | 430 | 480 | 570 | 550 | 510 | 440 | 490 | 550 | 590 | 590 |

Calculate the mean count rate.

Calculate the standard deviation for the series of readings.

Calculate the detection limit as the concentration which would apparently give a reading equal to three times the standard deviation.

(Answers: 519 cps; 56 cps; 0.022 ng ml$^{-1}$)

Comment on possible reasons for the very noisy signals obtained and suggest how a lower detection limit might be obtained.

## B.2 LABORATORY EXERCISES

Before carrying out any of the following laboratory exercises, the appropriate safety audits should be performed. In particular, attention is drawn to the Control of Substances Hazardous to Health (COSHH) regulations.

## B.3 OPERATION AND OPTIMIZATION OF AN ATOMIC ABSORPTION SPECTROMETER AND DETERMINATION OF MAGNESIUM IN SYNTHETIC HUMAN URINE

For a given spectrometer it will be necessary to refer to the manufacturer's handbook and instructions for operation.

### B.3.1 Introduction

Practically all magnesium in the human body is found intracellularly (98%), with bones, muscles and the liver being the main sites of deposition. Of the magnesium ingested (10–20 mmol per day), two thirds is excreted in the stool with the remainder being reabsorbed, the majority of which is excreted through the kidneys which regulate magnesium concentration.

The biological activity of magnesium is manifold. It serves as an indicator for many enzyme reactions, predominantly phosphate metabolism (e.g. as a co-factor in the phosphorylation of glucose). It also plays a part in the regulation of the neuromuscular excitation process.

Atomic absorption spectrometry is the method of choice for clinical chemists, with serum usually being analysed to determine magnesium concentration, although in this case synthetic urine is being analysed to assess the amount of magnesium being processed by the kidneys.

### B.3.2 Objectives

1 To optimize the instrumental conditions for the analysis of magnesium by atomic absorption spectrometry.

2 To quantify magnesium in synthetic human urine using calibration standards.

## B.3.3  Instrumentation and reagents

1  Atomic absorption spectrometer with air–acetylene burner head. Pressurized acetylene cylinder. Air compressor.

2  Magnesium stock solution — using distilled or deionized water prepare six magnesium standards in 100 ml volumetric flasks of 0.1, 0.3, 0.4, 0.5, 0.6, and 0.8 µg ml$^{-1}$ concentration from the stock solution. Also prepare 250 ml of a 0.2 µg ml$^{-1}$ magnesium standard.

3  Synthetic human urine sample.

## B.3.4  Procedure

NOTE: Solution (either water or sample) must be continuously aspirated into the flame. Therefore, ensure that the sample delivery tube is immersed in solution at all times.

Ensure that you have prepared all of your standards before proceeding with the experiment.

1  Switch the spectrometer on and allow to warm up for 10 min.

2  Ensure that an air–acetylene burner head is fitted.

3  Insert a magnesium hollow-cathode lamp.

4  Set the lamp to the recommended value.

5  Turn the monochromator wavelength setting to 285 nm using the coarse adjustment, then use the fine wavelength control to tune in to the line maximum at 285.2 nm.

6  Ensure that the light from the lamp passes over the slot of the burner, 1–2 cm above the burner top, by placing a white card on the burner top.

7  Switch on the air compressor; ensure that the pressure and flow rate are correctly set.

8  Turn on the main valve on the acetylene cylinder and ensure that the cylinder and outlet pressures and flow rate are correct. If the cylinder pressure is below the recommended value do not attempt to ignite the flame, as this will result in acetone being ignited in the flame. (Acetylene when stored under pressure is dissolved in acetone.)

9  Ignite the flame and ensure that it is blue with a very faint tinge of yellow above the primary cone. If it appears to be too fuel lean or fuel rich, adjust the fuel flow rate control accordingly.

**10**   After ignition **aspirate water continuously**. Using a 10 cm$^3$ measuring cylinder, check that the uptake rate of the nebulizer is within the range quoted by the manufacturer. Adjust if necessary.

**11**   Zero the absorbance reading.

**12**   Set the instrument integration time (or the read out update) for 2 s.

**13**   Aspirate the 0.2 µg ml$^{-1}$ standard. This should give an absorbance reading of about 0.2.

**14**   *Effect of hollow-cathode lamp current.* Vary the hollow-cathode lamp current in steps of 1 mA, ±3 mA either side of the manufacturer's recommended value, ensuring that the maximum allowable lamp current is not exceeded. Take five replicate absorbance readings at each lamp current setting. Remember to zero the instrument at each setting whilst aspirating a blank and adjust the gain if necessary before taking the five replicate readings. Calculate the average absorbance and relative standard deviation (RSD) for each setting and plot a graph of absorbance against lamp current. The RSD is defined as standard deviation ($s$) of the absorbances divided by the mean absorbance ($\bar{x}$). [RSD = $(s/\bar{x}) \times 100$]. Select the lamp current which gives the highest absorbance with an acceptable signal-to-noise ratio (indicated by the RSD).

**15**   *Effect of burner head height.* Raise the burner head to the highest position which will not obscure the path of the light beam. Zero the absorbance read out and obtain five replicate absorbance readings for the 0.2 µg ml$^{-1}$ magnesium standard. Lower the burner head in 2 mm stages and obtain five replicate absorbance readings for the 0.2 µg ml$^{-1}$ magnesium standard at each height. Calculate the mean absorbance and RSD for each setting and plot a graph of absorbance against burner head height. Select the burner head height which gives the highest absorbance with an acceptable signal to noise ratio (indicated by the RSD).

**16**   *Effect of varying the fuel/air ratio.* Vary the fuel/air ratio by increasing the fuel flow rate in steps until a strongly yellow (fuel rich) flame is obtained. Zero the absorbance read out and obtain five replicate absorbance readings for the 0.2 µg ml$^{-1}$ magnesium standard. Reduce the fuel flow rate in stages and obtain five replicate absorbance readings for the 0.2 µg ml$^{-1}$ magnesium standard at each setting. Calculate the mean absorbance and RSD for each setting and plot a graph of absorbance against fuel/air ratio. Select the fuel/air ratio which gives the highest absorbance with an acceptable signal-to-noise ratio (indicated by the RSD).

17 *Adjusting the nebulizer.* Measure the sample uptake rate using a 10 ml measuring cylinder and stop-watch. Report the result in ml min$^{-1}$. Compare this with your previous reading. Note that the sample uptake rate and flame conditions are interdependent and that adjusting one will require re-optimization of the other. Therefore, do not adjust the sample uptake rate once the fuel/air ratio has been optimized.

18 *Effect of monochromator slit width.* For a range of monochromator slit width settings between 0.2 and 2.0 nm, zero the absorbance read-out and obtain five replicate absorbance readings for the 0.2 μg ml$^{-1}$ magnesium standard. Calculate the mean absorbance and RSD for each setting and plot a graph of absorbance against monochromator slit width. Select the monochromator slit width setting which gives the highest absorbance with an acceptable signal-to-noise ratio (indicated by the RSD).

19 The instrument should now be optimized with regard to the above parameters.

20 *Determination of magnesium in synthetic human urine.* Using the calibration standards already prepared, obtain five replicate readings for each magnesium standard and calculate the mean absorbance and RSD for each standard. Using a statistical software package to construct a calibration graph which includes RSD error bars. Report the linear regression equation and the squared product moment correlation coefficient for the line of best fit. Identify the linear working range. Using a blank magnesium standard estimate the instrumental limit of detection [where the limit of detection is defined as the sum of the mean blank signal plus three standard deviations of the mean blank signal $(X_{lod} = \overline{X}_{bl} + 3s_{bl})$].

21 You are provided with a sample of synthetic urine. Dilute it 200× and aspirate. Obtain five replicate absorbance readings. Calculate the mean absorbance and RSD. Ensure that the absorbance reading lies in the linear response range of your calibration graph. If not, dilute appropriately. Use the regression equation to determine the concentration of magnesium in the sample.

22 Shut down the instrument using the manufacturer's recommended procedure.

## B.3.5 Discussion

Include answers to the following questions:

1 Explain the effect of adjusting each of the instrumental parameters on the quality of the analytical signal.

2  How does each of these parameters affect the accuracy and precision of the procedure.

3  Report the analytical figures of merit (i.e. limit of detection, linear range, characteristic concentration) for the magnesium calibration and compare with literature values.

4  Report (with RSD) the result for the concentration of magnesium in the synthetic urine sample and compare with the normal clinical range.

## B.3.6  References

1.  Skoog, D.A. and Leary, J.A. (1992) *Principles of Instrumental Analysis*, 4th edn, 543.08.SKO.
2.  Christian, G.D. (1986) *Analytical Chemistry*, 4th edn, 543.CHR.
3.  Alkemade, C.T.J. and Herrmann, R. (1979) *Fundamentals of Analytical Flame Spectroscopy*, Hilger, Bristol.
4.  Cresser, M. (1994) *Flame Spectrometry in Environmental Chemical Analysis: a Practical Guide*, Royal Society of Chemistry, Cambridge.

## B.4  DETERMINATION OF SODIUM IN SOIL EXTRACTS BY ATOMIC EMISSION SPECTROMETRY

For a given spectrometer it will be necessary to refer to the manufacturer's handbook and instructions for operation.

## B.4.1  Introduction

The major ions in surface and underground waters used for irrigation are $Ca^{2+}$, $Mg^{2+}$, $Na^+$, $Cl^-$, $SO_4^{2-}$ and $HCO_3^-$ with the relative proportion of sodium to calcium and magnesium having an important effect on the quality of irrigation water.

If the equilibrium between the exchangeable and solution cations changes so that adsorbed $Ca^{2+}$ and $Mg^{2+}$ ions are replaced by $Na^+$, then careful soil management will be required to avoid swelling and loss of permeability, and possibly deflocculation and translocation, when leached with good quality water.

The former process is reversible whereas the latter may lead to an irreversible deterioration of soil structure. Hence the determination of sodium in soil extracts is an important indicator of soil quality. A convenient method for the determination of sodium in soil extracts and other aqueous media is based on the characteristic spectra that sodium emits upon being aspirated into an air–acetylene flame.

## B.4.2 Objectives

1 To investigate the effect of potassium on the analysis of sodium by atomic emission spectrometry.

2 To quantify sodium in a soil extract using calibration standards.

## B.4.3 Instrumentation and reagents

1 Atomic absorption/atomic emission spectrometer with air–acetylene burner head. Pressurized acetylene cylinder. Air compressor.

2 Sodium stock solution.

3 Potassium stock solution.

4 Ammonium chloride (1 mol l$^{-1}$).

5 Soil extract.

6 Using distilled or de-ionized water prepare six standards in 100 ml volumetric flasks containing 10 µg ml$^{-1}$ sodium and 500, 100, 50, 25, 10 and 0 µg ml$^{-1}$ potassium. Also prepare a 250 ml standard containing 10 µg ml$^{-1}$ sodium and 1000 µg ml$^{-1}$ potassium.

7 Using distilled or de-ionized water prepare five standards in 100 ml volumetric flasks containing 0.5 mol l$^{-1}$ ammonium chloride, 1000 µg ml$^{-1}$ potassium and 10, 8, 5, 2 and 0 µg ml$^{-1}$ sodium.

8 Using distilled or de-ionized water prepare a sample for analysis by pipetting 25 ml of the unknown soil extract into a 50 ml volumetric flask and adding an appropriate volume of potassium stock solution to give a final concentration of 1000 µg ml$^{-1}$.

## B.4.4 Procedure

NOTE: Solution (either water or sample) must be continuously aspirated into the flame. Therefore, ensure that the sample delivery tube is immersed in solution at all times. Rinse all glassware with distilled or de-ionized water before preparing standards to avoid sodium contamination. Ensure that you have prepared all of your standards before proceeding with the experiment.

1 Switch the spectrometer on and allow to warm up for 10 min.

2 Ensure that an air–acetylene burner head is fitted.

3   Turn the monochromator wavelength setting to 589 nm.

4   Set the slit width to 0.15 nm.

5   Ensure that the instrumental mode switch is set to flame emission.

6   Switch on the air compressor; ensure that the pressure and flow rate is correctly set.

7   Turn on the main valve on the acetylene cylinder and ensure that the cylinder and outlet pressures and flow rate are correct. If the cylinder pressure is below the recommended value do not attempt to ignite the flame, as this will result in acetone being ignited in the flame. (Acetylene when stored under pressure is dissolved in acetone.)

8   After ignition **aspirate water continuously**. Using a 10 cm³ measuring cylinder, check that the uptake rate of the nebulizer is within the range quoted by the manufacturer. Adjust if necessary.

9   Ensure that the flame is blue with a very faint tinge of yellow above the primary cone. If it appears to be too fuel lean or fuel rich, adjust the fuel flow rate control carefully.

10  After ignition **aspirate water continuously**.

11  Zero the emission reading.

12  Using the top standard (10 µg ml$^{-1}$ sodium and 1000 µg ml$^{-1}$ potassium), fine tune the monochromator setting to achieve maximum emission signal. Set the voltage of the photo multiplier tube (PMT) until a reading of about 90 is obtained with this solution, then increase the scale expansion control until a reasonable signal is achieved. The increase in signal is offset by an increase in noise.

13  Set the instrument integration time control to 1 s.

14  *The effect of potassium.* Take five replicate emission readings of each of the six potassium standards. Remember to zero the instrument between each standard while aspirating a blank. Calculate the mean emission and RSD for each standard and plot a graph of sodium emission intensity against potassium concentration. The RSD is defined as standard deviation ($s$) of the emission intensities divided by the mean emission intensity ($\bar{x}$)[RSD $= (s/\bar{x}) \times 100$].

15  *Determination of sodium in a soil extract.* Using the five sodium calibration standards already prepared obtain five replicate readings for each sodium standard and calculate the mean emission intensity and RSD for each standard. Aspirate the standards in ascending and descending order of concentration. Aspirate a blank between each standard and zero the instrument. Using a statistical software package construct a

calibration graph of the pooled data (ascending and descending) which includes RSD error bars. Identify the linear working range. Report the linear regression equation and the squared product moment correlation coefficient for the line of best fit. Using a paired $t$-test determine whether there is any significant difference between the emission data of the standards introduced in ascending and descending order. The significant difference $t$ is defined as $t = \bar{x}_d\sqrt{n/s_d}$, where $\bar{x}_d$ is the mean difference between equivalent standards in ascending and descending order, $n$ is the number of standards and $s_d$ the standard deviation of the differences. The critical value of $|t|$ for a confidence interval of 95% is 2.78. Using a blank sodium standard, estimate the instrumental limit of detection [where the limit of detection is defined as the sum of the mean blank signal plus three standard deviations of the mean blank signal $(X_{lod} = \bar{X}_{bl} + 3s_{bl})$]. Obtain five replicate emission intensity readings of the prepared sample solution. Calculate the mean emission intensity and RSD. Ensure that the emission reading lies in the linear response range of your calibration graph. If it does not, dilute appropriately. Use the regression equation to determine the concentration of sodium in the sample.

16  Shut down the instrument using the manufacturer's recommended procedure.

## B.4.5  Discussion

Include answers to the following questions:

1  Explain the shape of the graph showing the effect of adding potassium to sodium. What is the mechanism of this effect and why is potassium added to all solutions in the determination of the sodium in the extract? Would you expect to see the same effects in atomic absorption and, if so, why?

2  Report the analytical figures of merit (i.e. limit of detection, linear range) for the sodium calibration and compare with literature values. Discuss the implications of using a curved calibration graph.

3  Report (with RSD) the result for the concentration of sodium in the soil extract sample and compare with the literature values for soils.

4  What are the most probable sources of error in this experiment? Does the method used offer an acceptable level of precision and accuracy for the determination of sodium in soil extracts and what are the reasons for your conclusions?

5  Comment on the results of the paired $t$-test.

## B.4.6 References

1.  Skoog, D.A. and Leary, J.A. (1992) *Principles of Instrumental Analysis*, 4th edn, Saunders College Publishing, Orlando, Florida, USA.
2.  Christian, G.D. (1994) *Analytical Chemistry*, 5th edn, Wiley, New York.
3.  Alkemade, C.T.J. and Herrmann, R. (1979) *Fundamentals of Analytical Flame Spectroscopy*, Hilger, Bristol.
4.  Cresser, M. (1994) *Flame Spectrometry in Environmental Chemical Analysis: a Practical Guide*, Royal Society of Chemistry, Cambridge.
5.  Miller, J. C. and Miller, J.N. (1993) *Statistics for Analytical Chemistry*, 3rd edn, Ellis Horwood, Chichester.

# B.5 GRAPHITE FURNACE ATOMIC ABSORPTION SPECTROMETRY

## B.5.1 Objectives

Optimization of drying, ash and atomization temperatures; calibration and determination of Cu.

## B.5.2 Introduction

The technique is prone to very severe interference effects unless certain precautions are taken. The sample is dried, ashed and atomized according to the temperatures programmed into the instrument. Each stage has its own potential problems, and therefore each stage must be optimized. If the drying stage is too hot, the sample may boil vigorously and splatter across the tube, causing problems with precision. If too low a temperature is used, the time required for drying increases to such an extent that the entire programme is prohibitively long. The temperature used should be approximately 20 °C higher than the boiling point of the solvent, but in any case should lead to steady, even drying of the sample. This can be observed by the use of a mirror.

If the ash stage temperature is incorrectly set, several problems may arise. If it is too low, the matrix (i.e. the components in the sample that can lead to interferences) is not removed and hence they obscure the light beam when atomic absorption is being measured. If it is too high, then the analyte may be lost by volatilization as a salt (often chlorides) or by atomization. It is therefore necessary to construct an 'ash curve' to determine the optimum ash temperature.

The atomize temperature should also be optimized. If too low a temperature is used, the analyte will not atomize and hence no signal will be obtained. If it is too high, then the tube will experience unnecessary wear and its lifetime will be shortened. An added problem for many analytes is that they may react with the graphite and form extremely refractory carbides, which will also decrease the atomic absorption signal. The optimum atomize temperature may be found by constructing an 'atomize curve'.

## B.5.3 Procedure

1 First, optimize the temperature programme. In the first instance, only the dry temperature need be optimized, and so aliquots of water can be dispensed into the tube. A temperature range of 100–140 °C can be tried, and the temperature that gives complete drying without frothing or spitting selected.

2 Prepare a series of copper standards ranging from 0 to 100 ng ml$^{-1}$, in 2% nitric acid.

3 Set the ash temperature to 400 °C for 10 s, which should be low enough to prevent any losses of copper before the atomize stage, and set the atomize temperature to 1700 °C for 5 s. Set the tube-clean cycle to 2500 °C for 5 s.

4 Deliver 10 µl of a standard solution into the graphite furnace and run the furnace program.

5 Repeat the above procedure while increasing the atomization temperature to 2300 °C in stages of 100 °C. Plotting a graph of peak area against atomize temperature will yield a sigmoid-shaped curve. The optimum atomize temperature is that which is just on the higher plateau region.

6 Once the optimum drying and atomize temperatures have been selected, the optimum ash temperature can be determined. Select the optimum dry and atomize temperatures and an ash temperature of 400 °C. Note the peak area produced at atomization, and then repeat the experiment increasing the ash temperature to 1200 °C in 200 °C steps. A graph of ash temperature against signal should then be plotted. The optimum ash temperature is that just before the signal starts to decrease.

7 Using the optimum temperature programme, prepare a calibration curve.

8 Take a portion of deionized, distilled water and get a volunteer to dip his/her finger into it for 15 s. Determine the concentration of copper in this sample and a sample of tap-water.

## B.5.4  Discussion

1  Are there any particular problems associated with Cu determination by GFAAS?

2  If the sample was a seawater matrix would you anticipate any problems with the analysis?

3  What methods could you use to overcome matrix interferences?

## B.5.5  Reference

1.  Haswell, S. (Ed.) *Atomic Absorption Spectrometry*. Elsevier, New York, 1991.

## B.6  SPECIATION OF ARSENIC COMPOUNDS BY ION-EXCHANGE HIGH-PERFORMANCE LIQUID CHROMATOGRAPHY WITH HYDRIDE GENERATION ATOMIC FLUORESCENCE DETECTION

### B.6.1  Objectives

Speciation of a solution of mixed arsenic standards; determine the limit of detection; observe and explain the effects of pre-reduction.

### B.6.2  Instrumentation and reagents

### B.6.2.1  High-performance liquid chromatography

*Mobile phase:*  Solutions of 0.1 and 0.0001 M potassium sulphate ($K_2SO_4$) (Aristar, BDH) freshly prepared and buffered to pH 10 with a few drops of ammonia solution (AnalaR, BDH).

*Pump:*  HPLC reciprocating pump (Waters) with a Rheodyne switching valve.

*Column:*  300 mm × 4.6 mm i.d. column packed with Benson BA-X10 strong ion-exchange packing material, placed in a water-bath at 58 °C.

The solution of 0.0001 M $K_2SO_4$ should be pumped through the column at 0.5 ml min$^{-1}$ for 30 min to allow the column to equilibrate. After this time, raise the flow rate of the mobile phase to 1.2 ml min$^{-1}$.

A summary of the HPLC conditions is given in Table B.1.

**Table B.1**  HPLC conditions.

| | |
|---|---|
| Mobile phase 1 | 0.0001 M $K_2SO_4$ (pH 10) |
| Mobile phase 2 | 0.1 M $K_2SO_4$ (pH 10) |
| Column | Strong ion exchange |
| Column size | 300 mm × 4.6 mm i.d. |
| Column temperature | 58 °C |
| Column packing material | Benson BA-X10 |
| Mobile phase flow rate | 1.2 ml min$^{-1}$ |
| Sample loop | 200 µl |

## B.6.2.2  Hydride generation atomic fluorescence

*Hydride generator:*  PS Analytical continuous flow hydride generator.

*Atomic fluorescence spectrometer:*  PS Analytical Excalibur atomic fluorescence spectrometer (AFS).

*Reductant:*  A solution of 1% w/v sodium borohydride (NaBH$_4$) should be freshly prepared in 0.1 M sodium hydroxide (NaOH), and placed in the 'reductant' reagent compartment of the PS Analytical continuous flow hydride generator.

*Acid blank:*  A solution of 3 M hydrochloric acid should be prepared and placed in the 'acid blank' reagent compartment of the AFS.

The inlet line to the gas–liquid separator should be removed and connected to a plastic T-piece, which is also connected to the outlet from the HPLC column. The T-piece should then be connected to the gas–liquid separator with 7 cm of Teflon tubing.

When the hydride generation apparatus is then switched on, the NaBH$_4$ and HCl solutions are pumped, via a peristaltic pump, to the T-piece where they are mixed with the column eluent before entering the gas–liquid separator. The gases formed in the separator are swept, through a membrane drier, into the AFS using an argon carrier flow. The AFS instrument should be connected to a chromatographic integrator.

The instrumental conditions for the hydride generation atomic fluorescence instrument are given in Table B.2. A diagram of the instrumental set-up is shown in Fig. B.1.

## B.6.3  Procedure

1  Prepare a 100 ng ml$^{-1}$ mixed standard solutions of arsenic by serial dilution of 1000 µg ml$^{-1}$ stock solutions of dimethylarsenic (DMA), monomethylarsenic (MMA), arsenic(III) and arsenic(V)

**Table B.2** Instrumental conditions for hydride generation atomic fluorescence spectrometry.

| | |
|---|---|
| Carrier gas flow rate | 250 ml min$^{-1}$ |
| Membrane drier purge gas | 500 ml min$^{-1}$ |
| Reductant | 1% NaBH$_4$ in 0.1 M NaOH |
| Acid blank | 3 M HCl |
| Lamp | Boosted arsenic hallow-cathode lamp |
| Lamp primary current | 27 mA |
| Lamp boost current | 35 mA |

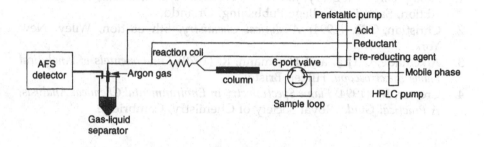

**Figure B.1**

2　Inject the standard into the sample loop of the HPLC instrument and observe the output of the atomic fluorescence instrument on the chromatographic integrator. On elution of the first arsenic species, switch the mobile phase from 0.0001 M K$_2$SO$_4$ to 0.1 M K$_2$SO$_4$ (step gradient elution).

3　Switch back to the 0.0001 M K$_2$SO$_4$ mobile phase and allow the column to re-equilibrate for 15 min after the chromatographic run (each run lasts about 25 min).

4　Insert a 10 m long Teflon reaction coil between the HPLC column and the gas–liquid separator with a Teflon T-piece between the reaction coil and the HPLC column, as shown in Fig. B.1.

5　Pump a 0.7% solution of L-cysteine in 0.05 M HCl at a rate of 0.2 ml min$^{-1}$ into the T-piece.

6　Repeat steps 2 and 3 above.

## B.6.4　Discussion

1　What were the main advantages of using AFS as the method of detection for this analysis?

2 Why is it necessary to perform pre-reduction before hydride generation?

3 This method can be extended to speciate other forms of arsenic such as arsenobetaine. What problems would you forsee using this method to determine arsenobetaine, and what modifications to the method could you perform to overcome them.

## B.6.5 References

1. Skoog, D.A. and Leary, J.A. (1992) *Principles of Instrumental Analysis*, 4th edition, Saunders College Publishing, Orlando.
2. Christian, G.D. (1994) *Analytical Chemistry*, 5th edition, Wiley, New York.
3. Alkemade, C.Th.J. and Herrmann, R. (1979) *Fundamentals of Analytical Flame Spectroscopy*, Hilger, Bristol.
4. Cresser, M. (1994) *Flame Spectrometry in Environmental Chemical Analysis. A Practical Guide*, Royal Society of Chemistry, Cambridge.

## B.7 INTRODUCTION TO ISOTOPE DILUTION INDUCTIVELY COUPLED PLASMA MASS SPECTROMETRY

This practical exercise makes use of an istopically enriched cadmium standard which is available from a number of sources. However, any element with an enriched isotope can be used in place of cadmium, provided that the natural abundance is known.

## B.7.1 Objectives

Light-up, ion lens tuning, mass calibration, optimization of data acquisition parameters, determination of Cd in water by isotope dilution

## B.7.2 Instrumentation and reagents

1 Enriched cadmium isotopic standard (Spectrascan Stable Isotope Solution, Teknolab A/S, P.O. Box 131, N-1441 Drobak, Norway).
2 Polypropylene bottles.
3 Pipettes.
4 Mass calibration tuning solution.
5 Torch, spray chamber and cones.

## B.7.3 Theory

Quadrupole mass spectrometers and their associated ion optics do not transmit ions of different mass equally. In other words, if an elemental solution composed of two isotopes with an exactly 1:1 ratio is analysed using ICP-MS, then a 1:1 isotopic ratio will not necessarily be observed. In practice, transmission through the quadrupole increases up to the mid-mass range (ca $m/z$ 120), then levels off or decreases gradually up to $m/z$ 255. This so-called mass bias will differ depending on mass, with the greatest effects occurring at low mass, the least effect in the mid-mass range and intermediate effects at high mass. Even very small mass biases can have deleterious effects on the accuracy of isotope ratio determinations, so a correction must always be made using an isotopic standard of known composition, as shown in Eqn. B.1.

$$C = \frac{R_t}{R_o} \tag{B.1}$$

where $C$ = mass bias correction factor, $R_t$ = true isotopic ratio for the isotope pair and $R_o$ = observed isotope ratio for the isotope pair. This correction factor will be applied to the isotope ratio determined for the sample.

### B.7.3.1 Sample preparation

To perform the isotope dilution analysis we will use an enriched Cd solution which has the isotopic abundances shown in Table B.3. We will use the $^{106}Cd$:$^{111}Cd$ isotopic pair for the analysis.

To achieve best accuracy and precision it is necessary to spike the sample solution with the enriched Cd to obtain a ratio as close as possible to unity.

**Table B.3** Isotopic abundances of enriched Cd solution.

| | Isotopic abundance (atom %) | |
|---|---|---|
| Isotope | Enriched Cd | Natural Cd |
| 106 | 79.013 ± 0.05 | 1.25 |
| 108 | 0.68 ± 0.01 | 0.9 |
| 110 | 3.03 + 0.02 | 12.5 |
| 111 | 2.60 ± 0.01 | 12.8 |
| 112 | 5.56 ± 0.01 | 24.1 |
| 113 | 1.73 ± 0.01 | 12.2 |
| 114 | 6.21 ± 0.01 | 28.7 |
| 116 | 1.18 ± 0.01 | 7.5 |

The requisite weight of the spike required to do this can be calculated using the equation.

$$R = \frac{A_x C_x W_x + A_s C_s W_s}{B_x C_x W_x + B_s C_s W_s} \tag{B.2}$$

where $R$ = isotope ratio of A to B, $A_x$ = atom fraction of isotope A ($^{106}$Cd) in sample, $B_x$ = atom fraction of isotope B ($^{111}$Cd) in sample, $A_s$ = atom fraction of isotope A ($^{106}$Cd) in spike, $B_s$ = atom fraction of isotope B ($^{111}$Cd) in spike, $W_x$ = weight of sample, $W_s$ = weight of spike, $C_x$ = concentration of element in sample and $C_s$ = concentration of element in spike.

The equation can be rearranged and the appropriate values substituted to calculate $W_s$:

$$W_s = \frac{W_x C_x (A_x - RB_x)}{C_s (RB_s - A_s)} \tag{B.3}$$

### B.7.3.2  Calculation of unknown

Equation B.2 can also be rearranged into the appropriate form to calculate $C_x$ as follows:

$$C_x = \frac{C_s W_s}{W_x} \times \frac{A_s - RB_s}{RB_x - A_x} \tag{B.4}$$

## B.7.4  Instrument start-up

A demonstration of the correct procedure for the ICP-MS instrument will be given. This will be an opportunity to get to know the instrument by discussions with the demonstrator and hands-on 'knob twiddling'. The following aspects will be covered.

### B.7.4.1  Preparation

A complex instrument such as ICP-MS requires careful preparation before use if reliable results are to be obtained. Useful preparative steps include:

- acid washing of spray chamber, torch and other glassware;
- cleaning of sampler and skimmer cones;
- preparation and alignment.

### B.7.4.2  Ignition

This procedure is automated in most modern instruments; however, it is as well to be aware of things that can go wrong beforehand in order to prevent possible damage. Areas of concern include:

- plasma formation without melting the torch;
- reflected power;
- punching the plasma;
- operating pressures in the mass spectrometer.

### B.7.4.3 Tuning for maximum signal

The initial aim is to optimize the system in order to obtain the maximum signal for a given concentration of analyte. A typical tuning solution comprises Be, Mg, Co, In, Ba, Pb and U at a concentration of 10 ng ml$^{-1}$.

### B.7.4.4 Mass calibration

The correct procedure for mass calibration will be demonstrated.

## B.7.5 Optimization

Before performing the actual isotope dilution analysis it is important to optimize the instrument to ensure precise and accurate data. The following aspects must be considered.

### B.7.5.1 Data acquisition parameters

The precision of any isotope ratio measurement depends to a great extent on the mass spectrometer operating conditions. Because a quadrupole mass spectrometer is a rapid sequential analyser, the frequency at which it switches between the masses to be ratioed and the amount of time spent collecting data at each mass must be optimized.

Two main modes of operation are possible. In the scanning mode, the whole profile of each of the masses is scanned. In the peak-hopping mode only several points (i.e. channels) over the peak profile are included in the data acquisition. The latter method is more rapid so is generally preferred, and is the method used here.

To optimize the instrument we will use a 10 ng ml$^{-1}$ solution of Cd which has been blended from natural Cd and the enriched $^{106}$Cd standard, to give a $^{106}$Cd: $^{111}$Cd ratio of approximately 1. Prepare this solution as follows:

1  Accurately weigh 100 µl of a 10 µg g$^{-1}$ natural Cd solution into a polypropylene bottle.

2  Calculate the weight of a 1 μg g$^{-1}$ solution of enriched $^{106}$Cd that you would need to add to the natural Cd to give a $^{106}$Cd:$^{111}$Cd ratio of approximately 1, using Eqn. B.3. Accurately weigh as close as possible to this mass of the enriched $^{106}$Cd solution into the polypropylene bottle.

3  Make up to 100 g with distilled, de-ionized water.

4  This is the 'optimization solution'.

### B.7.5.2  Dwell time

The dwell time is the time spent acquiring data at each of the channels which make up a peak in the mass spectrum. The length of time is measured in fractions of a millisecond, and will ultimately affect the frequency with which data is acquired at each mass. This will have a bearing on the final precision of the isotope ratio because of the influence of various sources of 'noise' on the analytical signal.

To optimize the dwell time, set up a data acquisition procedure using the dwell times shown in Table B.4, aspirate the 'optimization solution' and acquire count rate data for the $^{106}$Cd and $^{111}$Cd isotopes. The method for setting the data acquisition parameters will vary between instruments, but a similar procedure should be possible for all makes of instrument. Record the data in Table B.4, calculate the mean and RSD for the $^{106}$Cd:$^{111}$Cd ratio for each dwell time and hence determine the best precision.

### B.7.5.3  Points per peak

The number of points, or channels, which are chosen for each isotopic peak will also affect the frequency of data acquisition, and hence the final precision of the isotope ratio measurement. Three points per peak is a typical value, but this too could be optimized.

**Table B.4**  Dwell times and precision.

| Dwell time (ms) | Replicates | Mean $^{106}$Cd:$^{111}$Cd ratio | SD | RSD (%) |
|---|---|---|---|---|
| 0.05 | 10 | | | |
| 0.1 | 10 | | | |
| 0.5 | 10 | | | |
| 1 | 10 | | | |
| 5 | 10 | | | |
| 10 | 10 | | | |
| 50 | 10 | | | |

**Table B.5**.

| Parameter | Mass bias solution | Sample |
|---|---|---|
| $R_o$ | | |
| $C$ | | |
| $R_t$ | | |
| $A_x$ | | |
| $B_x$ | | |
| $A_s$ | | |
| $B_s$ | | |
| $W_s$ | | |
| $W_s$ | | |
| $C_x$ | | |
| $C_s$ | | |

# B.7.6 Analysis of water sample

To aid your calculations, record your results in Table B.5, and use Eqns. B.1–B.4.

## B.7.6.1 Preparation of mass bias solution

1 Accurately weigh 100 μl of a 10 μg g$^{-1}$ natural Cd solution into a polypropylene bottle.

2 Calculate the weight of a 1 μg g$^{-1}$ solution of enriched $^{106}$Cd that you would need to add to the natural Cd to give a $^{106}$Cd:$^{111}$Cd ratio of approximately 1, using Eqn. B.3. Accurately weigh as close as possible to this mass of the enriched $^{106}$Cd solution into the polypropylene bottle.

3 Make up to 100 g with distilled, de-ionized water.

## B.7.6.2 Preparation of spiked sample solution

1 Prepare 250 ml of a 10 μg g$^{-1}$ natural Cd solution in a polypropylene bottle. This will be the synthetic water sample.

2 Accurately weigh 100 g of the synthetic water sample into a polypropylene bottle.

3 Accurately weigh 0.15 g of the 1 μg g$^{-1}$ $^{106}$Cd enriched solution into the polypropylene bottle. This should give you a $^{106}$Cd:$^{111}$Cd ratio of approximately 1, but you can calculate this using Eqn. B.3 to check.

### B.7.6.3  Analysis

Using the optimized data acquisition conditions, aspirate the mass bias solution, then the sample, then the mass bias solution again. Perform 10 replicate integrations on the $^{106}$Cd and $^{111}$Cd isotopes for each sample, and print out the results.

## B.7.7  Calculations

Record your results in Table B.5 and complete the following calculations:

1  Calculate the true $^{106}$Cd:$^{111}$Cd isotope ratio, $R_t$, for the mass bias solution, using Eqn. B.2.

2  Calculate the observed $^{106}$Cd:$^{111}$Cd isotope ratio for the mass bias solution, $R_o$, for the two results which bracket the sample.

3  Using the mean of the results for $R_o$, calculate the mass bias correction factor, $C$, using Eqn. B.1.

4  Calculate the observed isotope ratio, $R_o$, for the sample, and multiply by the mass bias correction factor to obtain the true isotope ratio, $R_t$. Hence, calculate $C_x$ for the sample using Eqn. B.4.

## B.7.8  Discussion

1  Discuss the advantages and disadvantages associated with isotope dilution analysis.

2  What flaws can you identify in the method you have just followed?

3  Discuss the particular problems associated with the determination of Cd by isotope dilution analysis.

## B.7.9  References

1.  Fasset, J.D. and Paulsen, P.J., Isotope dilution mass spectrometry for accurate elemental analysis, *Anal. Chem.* 1989, **61**, 643A.

2.  Van Heuzen, A.A., Hoekstra, T. and van Wingerden, B., Precision and accuracy attainable with isotope dilution analysis applied to ICP-MS: theory and experiments, *J. Anal. At. Spectrom.* **4**, 483 (1989).

# APPENDIX C

# BIBLIOGRAPHY

A list of useful texts and reference works is appended here. No attempt has been made to give a selected bibliography of the many important original and review papers which have appeared. The advertising pages of *Analytical Chemistry* frequently contain readable introductory reviews of recent developments. Without doubt, the best tool for maintaining current awareness in this field is Atomic Spectrometry Updates published in the *Journal of Analytical Atomic Spectrometry*. These monthly reviews are produced by a distinguished international editorial board, and include comprehensive applications tables to enable practical workers to rapidly identify current methodology.

## C.1 JOURNALS

*Analyst*
*Analytica Chimica Acta*
*Analytical Chemistry*
*Analytical Communications*
*Analytical Letters*
*Analytical Sciences*
*Applied Spectroscopy*
*Atomic Spectroscopy*
*Microchemical Journal*
*Mikrochimica Acta*
*Spectrochimica Acta (Part B)*
*Talanta*
*Fresenius Journal of Analytical Chemistry*

Additionally, there are many papers published in non-English language journals, application-oriented journals (e.g. *Clinical Chemistry*) and conference proceedings. *Analytical Abstracts* and *Chemical Abstracts* will cover many of these. There are also the review journals *Progress in Analytical Spectroscopy, Trends in Analytical Chemistry* and *Spectrochimica Acta Reviews*.

## C.2 BOOKS

### C.2.1 General

*Flames. Their Structure, Radiation and Temperature*, Gaydon, A.G. and Wolfard, H.G., Chapman & Hall, London, 1979. See below.

*The Spectroscopy of Flames*, 2nd edn, Gaydon, A.G., Chapman & Hall, London, 1974. The classic fundamental text concerning the chemistry and spectroscopy of flames.

*Environmental Analysis using Chromatography Interfaced with Atomic Spectroscopy*, Harrison, R.M., and Rapsomanakis, S., Ellis Horwood, Chichester, 1989. Becoming dated, but still a valuable guide to this specialist area.

*Spectrochemical Analysis*, Ingle, J.D. and Crouch, S.R., Prentice Hall, Englewood Cliffs, NJ, 1988. A comprehensive treatment of atomic and molecular spectroscopy.

*Glow Discharge Spectroscopies*, Marcus, K. (Ed.), Plenum Press, New York, 1993. Covers the theory and application of glow discharges in depth.

*Statistics for Analytical Chemistry*, 3rd edn, Miller, J.C. and Miller, J.N., Ellis Horwood, Chichester, 1993. A text which no analytical chemist should be without, a concise and well written reference for the practising analytical chemist.

*Quality in the Analytical Chemistry Laboratory*, Pritchard, E. (Ed.), Wiley, Chichester, 1995. A concise practical guide to quality assurance procedures, this open learning text is ideal for beginners.

*Advances in Atomic Spectrometry*, Vols 1–3, Sneddon, J. (Ed.), JAI Press Greenwich, CT, 1997. Reviews of current status of atomic spectrometry; multi-authored so inconsistent in style and content.

*Quality Assurance of Chemical Measurements*, Taylor, J.K., Lewis, Chelsea, MI, 1987. A comprehensive guide to quality assurance.

### C.2.2 Atomic absorption and fluorescence spectrometry

*Fundamentals of Analytical Flame Spectroscopy*, Alkemade, C.T.J. and Herrmann, R., Hilger, Bristol, 1979. Dated, but a good account of the basic theory of flame spectroscopy.

*Solvent Extraction in Flame Spectroscopic Analysis,* Cresser, M.S., Butterworths, London, 1978. Now somewhat dated, but the methods are equally applicable for modern instrumentation.

*Flame Spectrometry in Environmental Chemical Analysis: A Practical Guide,* Cresser, M.S., Royal Society of Chemistry, Cambridge, 1995. A concise and very useful reference for the practising analyst.

*Atomic Absorption Spectrometry,* Haswell, S. (Ed.), Elsevier, New York, 1991. Theoretical and practical applications of AAS.

*Atomic Absorption and Fluorescence Spectroscopy,* Kirkbright, G.F. and Sargent, M., Academic Press, London, 1974. Now rather dated but nevertheless an unsurpassed treatment of theory.

## C.2.3  Plasma atomic emission spectrometry

*Line Coincidence Tables for Inductively Coupled Plasma Atomic Emission Spectrometry,* Boumans, P.W.J.M, Pergamon Press, New York, 1984. The most comprehensive compilation available of sensitive lines for use in ICP-AES, with listings of potential interferences.

*Inductively Coupled Plasma Emission Spectrometry, Parts I and II,* Boumans, P.W.J.M. (Ed.), Wiley, New York, 1987. A comprehensive account of the subject, with good chapters on theory, though now becoming dated.

*Inductively Coupled Plasmas in Analytical Atomic Spectrometry,* 2nd edn, Montaser, A., and Golightly, D.W. (Eds), VCH, New York, 1992. The standard text on the subject, an essential purchase for the serious student of inductively coupled plasma atomic emission spectrometry.

*An Atlas of Spectral Interferences in ICP Spectroscopy,* Parsons, M.L., Forster, A. and Anderson, D., Plenum Press, New York, 1980. Potential interferences from matrix elements.

*Automatic Atomic Emission Spectroscopy,* 2nd edn, Slickers, K., Brühlsche Universitätsdruckerei, Giessen, 1993. A very useful practical guide to arc and spark methods in the metallurgical industry.

*Inductively Coupled Plasma Atomic Emission Spectroscopy: An Atlas of Spectral Information,* Winge, R.K., Fassel, V.A., Peterson, V.J. and Floyd, W.A., Elsevier, New York, 1985. A compilation of sensitive lines for use in ICP-AES.

## C.2.4  Plasma mass spectrometry

*Inductively Coupled and Microwave Induced Plasma Sources for Mass Spectrometry,* Evans E.H., Giglio, J.J., Castillano, T.M. and Caruso, J.A., Royal Society of Chemistry, Cambridge, 1995. A tutorial review of both theory and practise, a useful, easy to understand primer, with particular emphasis on speciation.

*Plasma Source Mass Spectrometry*, Jarvis, K.E., Gray, A.L., Williams, J.G. and Jarvis, I., Royal Society of Chemistry, Cambridge, 1990.

*Handbook of Inductively Coupled Plasma Mass Specrometry*, Jarvis, K.E., Gray, A.L. and Houk, R.S., Blackie, Glasgow, 1992.

*Advances in Mass Spectrometry*, Todd, J. (Ed.), Wiley, New York, 1986. Covers a number of techniques including ICP-MS.

*Inorganic Mass Spectrometry*, Adams, F., Gijbels, R. and van Greken, R. (Eds), Wiley, New York, 1988. Covers a number of techniques including ICP-MS.

# INDEX

*Figure (f), Table (t)*